Library of Congress Cataloging-in-Publication

Eargle, John
Electroacoustical reference data / by John M. Eargle.
p. cm.
Includes bibliographical references and index.
ISBN 978-0-442-01397-4 ISBN 978-1-4615-2027-6 (eBook)
DOI 10.1007/978-1-4615-2027-6
1. Electro-acoustics--Charts, diagrams, etc. I. Title.
TK5981.E26 1994
621.382'021--dc20 94-2025
CIP

Originally published by Kluwer Academic Publishers in 2002

This printing is a digital duplication of the original edition.

Printed on acid-free paper.

ELECTROACOUSTICAL REFERENCE DATA

JOHN M. EARGLE

SPRINGER SCIENCE+BUSINESS MEDIA, LLC

CONTENTS

PREFACE

The need for a general collection of electroacoustical reference and design data in graphical form has been felt by acousticians and engineers for some time. This type of data can otherwise only be found in a collection of handbooks. Therefore, it is the author's intention that this book serve as a single source for many electroacoustical reference and system design requirements. In form, the volume closely resembles Frank Massa's *Acoustic Design Charts,* a handy book dating from 1942 that has long been out of print. The basic format of Massa's book has been followed here: For each entry, graphical data are presented on the right page, while text, examples, and references appear on the left page. In this manner, the user can solve a given problem without thumbing from one page to the next. All graphs and charts have been scaled for ease in data entry and reading.

The book is divided into the following sections:

A. General Acoustical Relationships. This section covers the behavior of sound transmission in reverberant and free fields, sound absorption and diffraction, and directional characteristics of basic sound radiators.

B. Loudspeakers. Loudspeakers are discussed in terms of basic relationships regarding cone excursion, sensitivity, efficiency, and directivity index, power ratings, and architectural layout.

C. Microphones. The topics in this section include microphone sensitivity and noise rating, analysis of directional properties, stereo microphone array characteristics, proximity effects, and boundary conditions.

D. Signal Transmission. This section covers many topics, including various pre-emphasis and de-emphasis characteristics, voltage and power relationships, losses in distribution systems, amplifier requirements, speech intelligibility requirements, and sound system equalization standards.

E. Psychoacoustical Data. Presented here are various loudness relationships, pitch and level relationships, stereophonic localization phenomena, and architectural acoustical considerations.

F. Musical Instruments. The basic characteristics of musical instruments are reviewed as they influence microphone choice and level settings in electroacoustical systems.

G. Analog Magnetic Recording. Track width standards, equalization standards, and various high-frequency losses are discussed in this section.

A comprehensive index helps the user identify specific graphs and relationships for the problem or project at hand.

PART ONE

GENERAL ACOUSTICAL RELATIONSHIPS

1. SOUND PRESSURE AND dB L_p (SOUND PRESSURE LEVEL)

In virtually all acoustical level measurements, a reference pressure of 20 $\mu N/m^2$ (20 micropascals) is used to determine dB L_p. The relevant equation is:

$$\text{dB } L_p \text{ (dB)} = 20 \log (P/P_0) \quad \ldots\ldots\ldots\ldots\ldots\ldots\ldots\ldots (1.1)$$

where $P_0 = 20\ \mu N/m^2$. An equivalent quantity in the CGS system of units is .0002 dynes/cm^2.

In Figure 1, there are two abscissa (x-axis) scales; the upper one refers to ordinate (y-axis) values at the left of the graph, and the lower one refers to values at the right.

EXAMPLE: Find the value of L_p corresponding to a pressure of 1 pascal.

Locate the value of 1 pascal on the lower abscissa and read upward until the bold line is intersected. At the right ordinate, read the value of 94 dB L_p.

EXAMPLE: Find the value of L_p corresponding to a pressure of 20 micropascals.

Locate the value of 20 micropascals on the upper abscissa. Note that it directly intersects the bold line. Reading at the left ordinate, we get 0 dB L_p.

Note: In much current literature, the term "dB-SPL" (sound pressure level) is used instead of the generally preferred dB L_p.

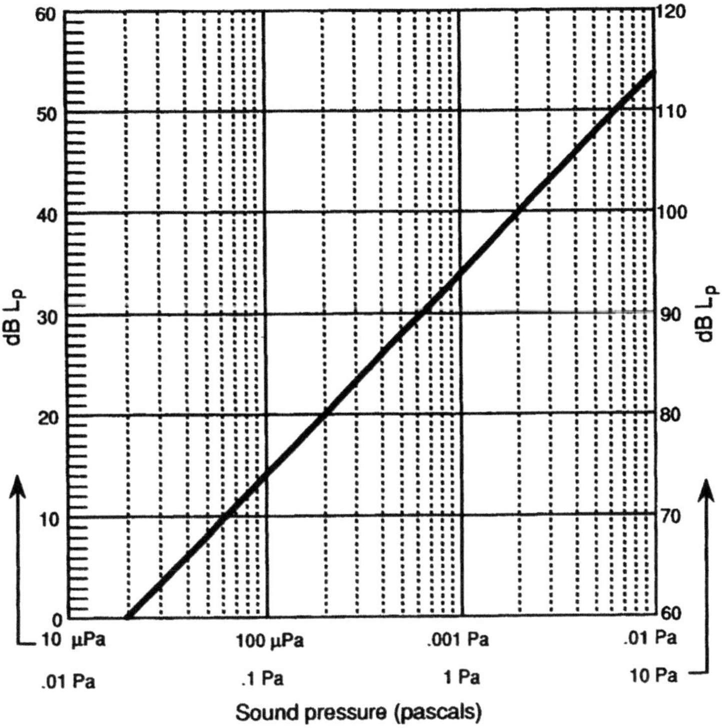

Figure 1. Sound Pressure and dB L_p (Sound Pressure Level)

2. Frequency and Wavelength in Air

In Figure 2, for frequencies from 10 Hz to 100 kHz, wavelength is plotted in meters (left ordinate) and in feet (right ordinate). Normal temperature is assumed. The relevant equations are:

wavelength (meters) = 339/frequency . . . . (2.1)

and

wavelength (feet) = 1130/frequency . . . . (2.2)

EXAMPLE: Find the wavelength corresponding to a frequency of 1 kHz.

Locate 1 kHz on the abscissa and read upward to bold line A. On the left ordinate, read a value of approximately .339. Moving on to bold line B, read the value of approximately 1.13 feet at the right ordinate.

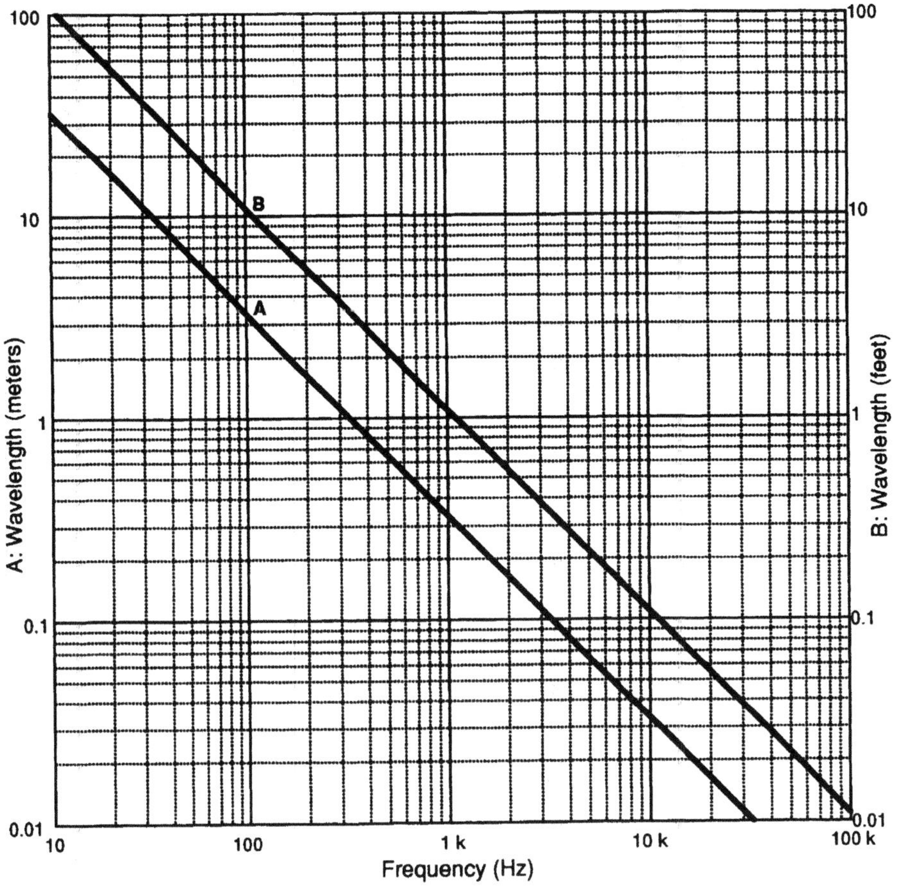

Figure 2. Frequency and Wavelength in Air

3. Inverse Square Losses in a Free Field

The nomograph presented in Figure 3 provides a simple way to determine by inspection the universe square level loss between any two measurement distances (meters or feet). The relevant equation is:

$$\text{Loss (dB)} = 20 \log (d/d_0) \quad (3.1)$$

where d_0 is equal to 1 meter or 1 foot, depending on the system of units.

EXAMPLE: Find the level difference under free-field conditions between a point source observed at distances of 10 and 36 meters.

Locate 10 at the bottom of the nomograph and read the value of 20 dB directly above it. Locate 36 at the bottom of the nomograph and read the value of 31 dB directly above it. Subtracting (36 – 20), we get the value of 16 dB.

EXAMPLE: Find the level difference under free-field conditions between a point source observed at a distance of 1 foot and 200 feet.

Locate 1 at the bottom of the nomograph and read the value of 0 dB directly above it. Locate 200 at the bottom of the nomograph and read the value of 46 dB directly above it. Subtracting (46 – 0), we get the value of 46 dB.

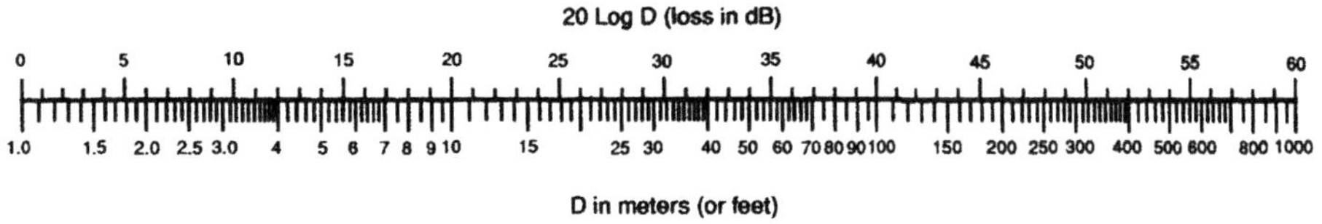

Figure 3. Inverse Square Losses in a Free Field

4. Attenuation with Distance from Plane and Line Sources in a Free Field

Typical inverse square loss is shown at Figure 4A, and the graph plots a loss of 6 dB per doubling of distance from the point source.

For an infinite line source, the attenuation with distance will be 3 dB per doubling of distance. For a line source of length A, the attenuation will be approximately 3 dB per doubling of distance measured out to a distance of A/π from the line source. Beyond the distance A/π, the loss approaches 6 dB per doubling of distance, as shown in Figure 4B.

Attenuation with distance from a plane source is shown at Figure 4C. Note that, at distances very close to the plane, there will be no attenuation with distance. This will continue out to a value of A/π, where A is the shorter side of the plane. Beyond this point, attenuation will be approximately 3 dB per doubling of distance out to a value of B/π, where B is the longer side of the plane. Beyond that point, the attenuation will approach 6 dB per doubling of distance.

When A and B are within a ratio of two or three of each other, it may be difficult to discern three clear regions of attenuation with distance (0, 3, and 6 dB per doubling of distance). Under these conditions, there will generally be a single broad transition range from point source to plane source. When A and B are in a ratio of about six or higher, then the three regions of attenuation slope become more apparent.

EXAMPLE: Determine the attenuation with distance from a 4 × 20 meter plane of uniform loudspeakers in a free field. Determine the values of $4/\pi$ and $20/\pi$ and plot them as shown at Figure 4D. The resulting graph then gives the approximate attenuation with distance.

In actual measurements, the attenuation plot with distance will not exhibit the sharp transitions shown in the graphs. Rather, the observed attenuation will be slightly less than those values.

References: Rathe (1969)
Gander and Eargle (1990)

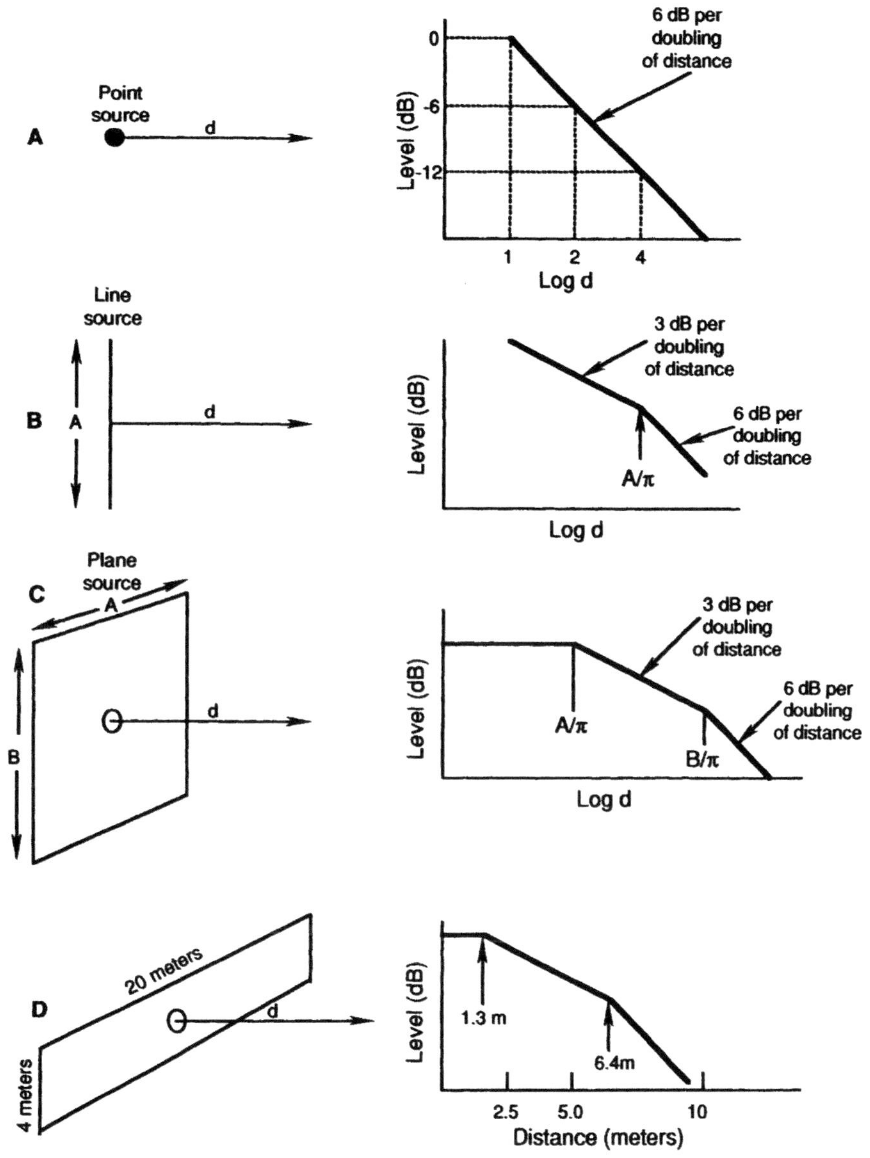

Figure 4. Attenuation with Distance from Plane and Line Sources in a Free Field

5. Atmospheric Sound Absorption as a Function of Frequency and Relative Humidity, I

Figure 5 shows the energy attenuation constant, m, in inverse meters (m^{-1}) as a function of frequency and relative humidity. In this form, the data can be used to determine the effects of atmospheric losses on reverberation time at high frequencies.

The Sabine reverberation time equation is:

$$T_{60} = \frac{.16V}{S\bar{\alpha} + 4mV} \quad \ldots\ldots (5.1)$$

where V is the room volume in cubic meters, $S\bar{\alpha}$ is the total boundary absorption, and m is the attenuation constant taken from the graph.

The Sabine equation in English units is:

$$T_{60} = \frac{.049V}{S\bar{\alpha} + 4mV} \quad \ldots\ldots (5.2)$$

EXAMPLE: Assume that a room has a volume of 5400 cubic meters and total boundary absorption of 1800 square meters at 4 kHz. At a relative humidity of 20%, determine the reverberation time.

The atmospheric term gives an additional 432 square meters of effective absorption: $4 \times (.02) \times (5400) = 432$. The reverberation time is thus:

$$T_{60} = 864/(1800 + 432) = .387 \text{ seconds}$$

If the atmospheric term had not been considered, the calculated reverberation time would be .48 seconds.

Reference: Knudsen and Harris (1950)

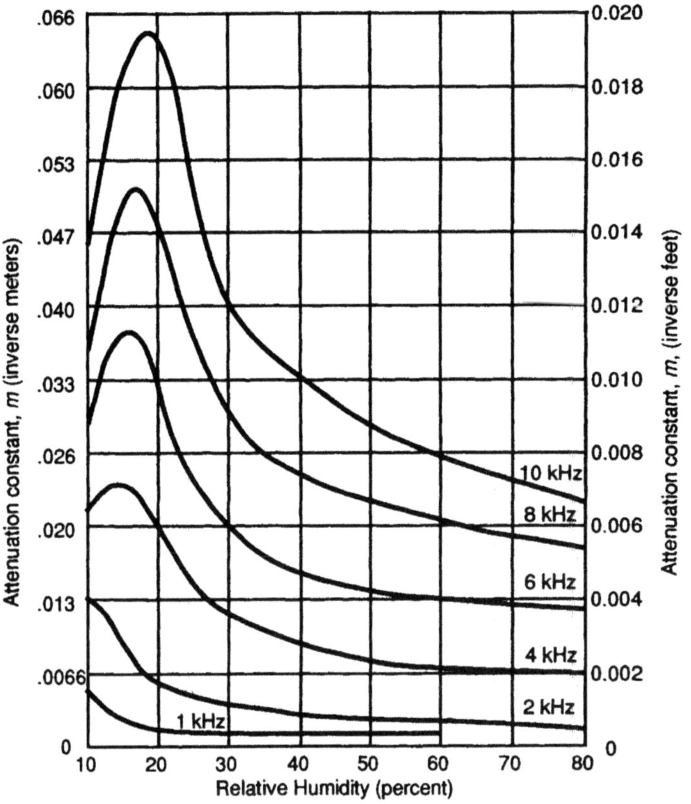

Figure 5. Atmospheric Sound Absorption as a Function of Frequency and Relative Humidity, I

6. Atmospheric Sound Absorption as a Function of Frequency and Relative Humidity, II

Essentially, the same data from section 5 are replotted in Figure 6, and extended to lower frequencies. The attenuation constant is given in dB per meter; in this form, it can be used to calculate the excess losses over distance due to atmospheric effects.

EXAMPLE: Assume that a sound source produces a level of 100 dB L_p at a distance of 1 meter. At a frequency of 10 kHz and at 20% relative humidity, what will be the level at a distance of 20 meters?

There will be both an inverse square loss term and an atmospheric loss term in the solution. First, calculate the inverse square term as 20 log 20 = 26 dB.

The atmospheric term is calculated as follows: Locate 10 kHz along the abscissa and move upward to the bold curve marked 20% RH. Read the value of approximately .35 at the intersection. Thus, the atmospheric term will be: (20) × (.35) = 7 dB, and the total loss will be: 26 + 7 = 33 dB.

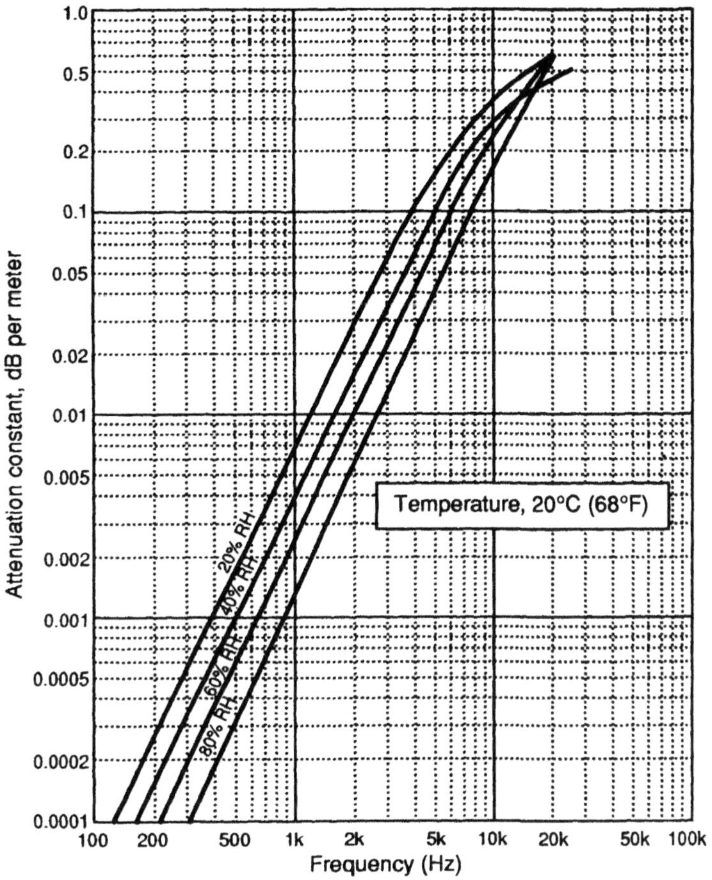

Figure 6. Atmospheric Sound Absorption as a Function of Frequency and Relative Humidity, II

7. Atmospheric Absorption Due to Inverse Square Losses and Relative Humidity

Figure 7 presents atmospheric attenuation data and inverse square losses at several distances for relative humidities of 20 and 80%. In this form, the data are useful for recording engineers in estimating sound pickup characteristics at large microphone distances.

EXAMPLE: An organ is to be recorded in a large space at an average distance of about 12 meters. How much high-frequency boost should be added if the relative humidity is about 20%?

By inspection, a loss commencing at 5 kHz, reaching a value of about 5 dB at 12 kHz, should be compensated for. As we will show later in section 89, the required difference can often be approximated by using a microphone designed for slightly peaked on-axis response.

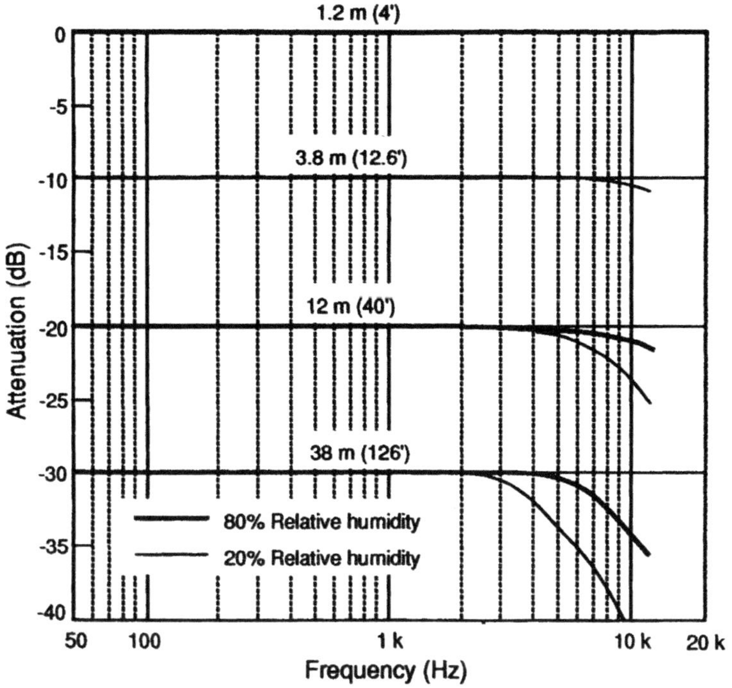

Figure 7. Atmospheric Absorption Due to Inverse Square Losses and Relative Humidity

8. NC AND PNC NOISE CRITERIA CURVES

Noise Criteria (NC) Curves

Noise criteria curves, developed in 1957, are used to determine the so-called NC rating, quantifying as a single number the noise present in an architectural space. Most measurements are made in octave-wide bands from 63 Hz to 8 kHz.

After the eight readings have been taken, they are plotted on a blank copy of the NC curves, shown at Figure 8A, and lines are normally drawn between the points. The NC rating for the space is generally stated as the lowest NC value that just touches the plotted points.

EXAMPLE: Assume that we have taken the following octave band noise readings in a space:

63 Hz	57 dB
125 Hz	49 dB
250 Hz	43 dB
500 Hz	37 dB
1000 Hz	32 dB
2000 Hz	28 dB
4000 Hz	25 dB
8000 Hz	24 dB

When these values are plotted on the NC template, we can see that the values hover around the NC-30 value. However, at 250 and 500 Hz, we can see that both values are 2 dB higher than the NC-30 curve in that region. Thus, we would have to make the assessment that the proper NC rating of the space would be NC-32, even though in most of the octave bands the noise levels were NC-30 or better.

It is accepted practice to average any three contiguous bands, provided that the highest of the three does not lie more than 2 dB above the average for the three.

Preferred Noise Criteria (PNC) Curves

Many acousticians have criticized the NC curves on the basis that a noise spectrum that is deliberately generated to match a particular noise criterion is apt to sound too "hissy and boomy," thus making it necessary to reduce noise levels at the low and high extremes by about 5 dB.

These objections have largely been met by the PNC curves proposed in 1971, which are shown at Figure 8B. They are used exactly the same way as the NC curves. Note that the curves appear straighter than the NC curves, indicating that the high- and low-frequency thresholds have been reduced.

Many acousticians feel that neither NC nor PNC ratings give a sufficient description of noise in critical environments used for sound recording. For further study, we recommend Cohen and Fielder (1992).

References: Beranek (1971)
Harris (1979)
Cohen and Fielder (1992)

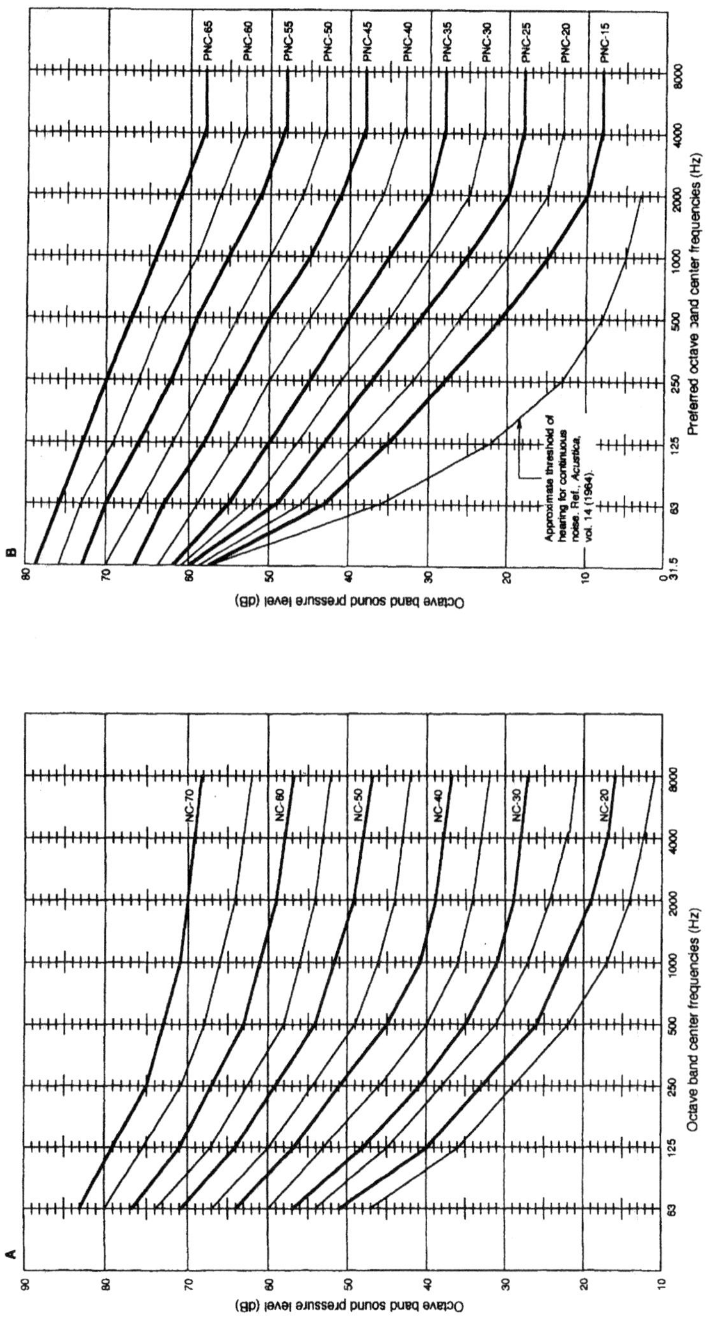

Figure 8. NC and PNC Noise Criteria Curves

9. Sound Transmission Class (STC) Curves

Sound transmission class curves, shown in Figure 9, are used to determine, as a single number, the effectiveness of walls and other architectural barriers in isolating airborne noise between spaces.

A given material, or a construction method involving a number of materials, may carry an NC rating. The NC rating is determined by constructing a barrier with that material between two suitably large rooms that are otherwise well isolated from one another. A wide-band noise source is placed in the room, and one-third octave readings are taken. Then, another set of one-third octave readings is made in the other room. Band-by-band differences are calculated; this gives the transmission loss (TL) between the two spaces.

The TL data are then plotted over the STC curve template. The TL data define the effective STC rating by the following rules: The average TL data must not exceed the target STC rating by 2 dB over the entire measurement band, and no individual TL value should exceed the STC contour by 8 dB.

References: Beranek (1971)
Harris (1979)

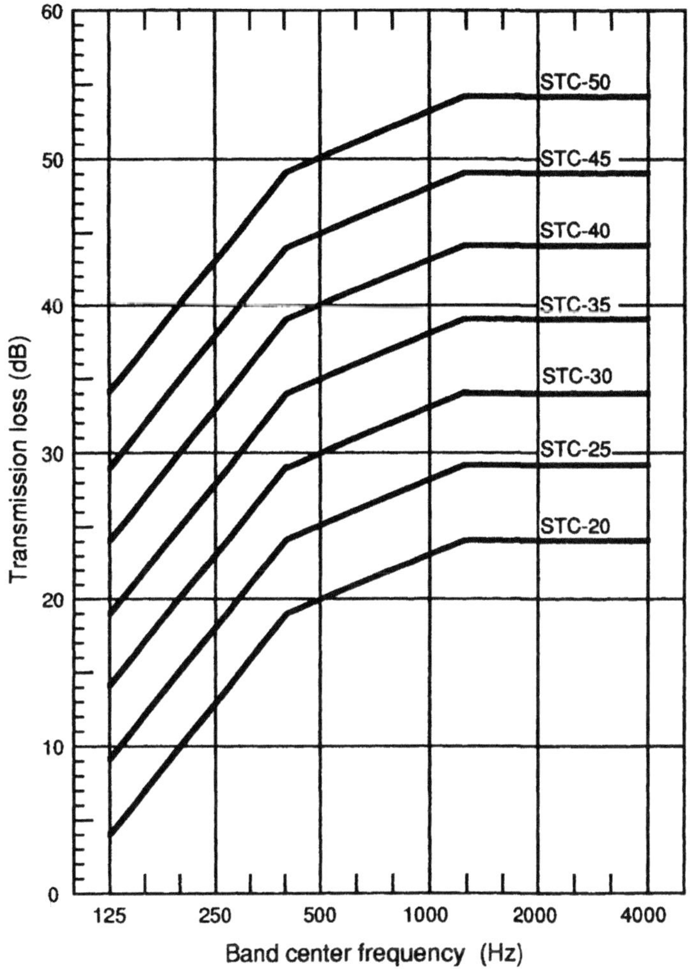

Figure 9. Sound Transmission Class (STC) Curves

10. Helmholtz Resonators

Helmholtz resonators provide a useful means of attenuating specific frequencies in certain architectural settings. The resonator consists of a specific enclosed volume coupled through a small orifice to the outside space. Resonance takes place when the system is excited by a frequency that is a function of the enclosed volume and the ratio of the volume and orifice diameter. The assumption is made that the thickness of the enclosure is negligible compared to the diameter of the orifice. Figure 10 shows the relation between the ratio of volume over orifice diameter to resonance frequency.

Resonance takes place at that frequency, f_0, where the springiness of the enclosed air and the mass of the air in the region of the orifice satisfy the resonance equation. The relevant equation is:

$$f_0 = c/2\pi[1.85a/V]^{1/2} \quad \text{. . . . .} \quad (10.1)$$

where c is the speed of sound (m/sec), a is the radius of orifice (mm), and V is the volume (liters).

EXAMPLE: What size orifice is required to construct a resonator that has a volume of 100 liters and resonates at 100 Hz?

First, find 100 Hz on the abscissa and move upward to the bold line. The intersection takes place at a value of 0.3 on the ordinate scale. This indicates that the ratio of volume divided by the orifice diameter should be equal to 0.3:

$$d = 100/0.3 = 333 \text{ mm}$$

Reference: Fletcher and Rossing (1991)

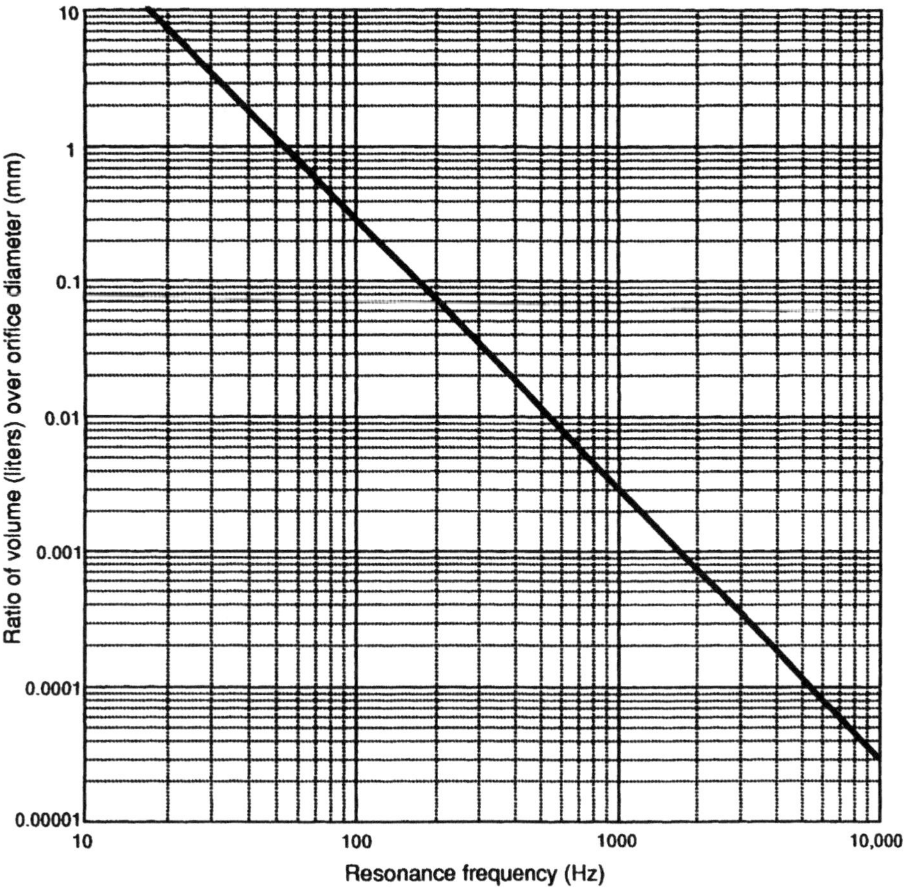

Figure 10. Helmholtz Resonators

11. Resonance Frequency for Pipes Open at Both Ends

Figure 11 gives the nominal length of an open pipe that will resonate at a particular frequency. This occurs when the pipe is approximately equal in length to one-half the wavelength of the frequency. Pipes that are particularly large in diameter may need to be modified in length, in order to determine their actual "speaking length."

EXAMPLE: Determine the nominal length of an open pipe that is to resonate at 250 Hz.

Locate 250 Hz along the abscissa and move upward to the bold line marked A. At that point, we read 0.7 along the left ordinate, indicating that the pipe will be 0.7 meters in length.

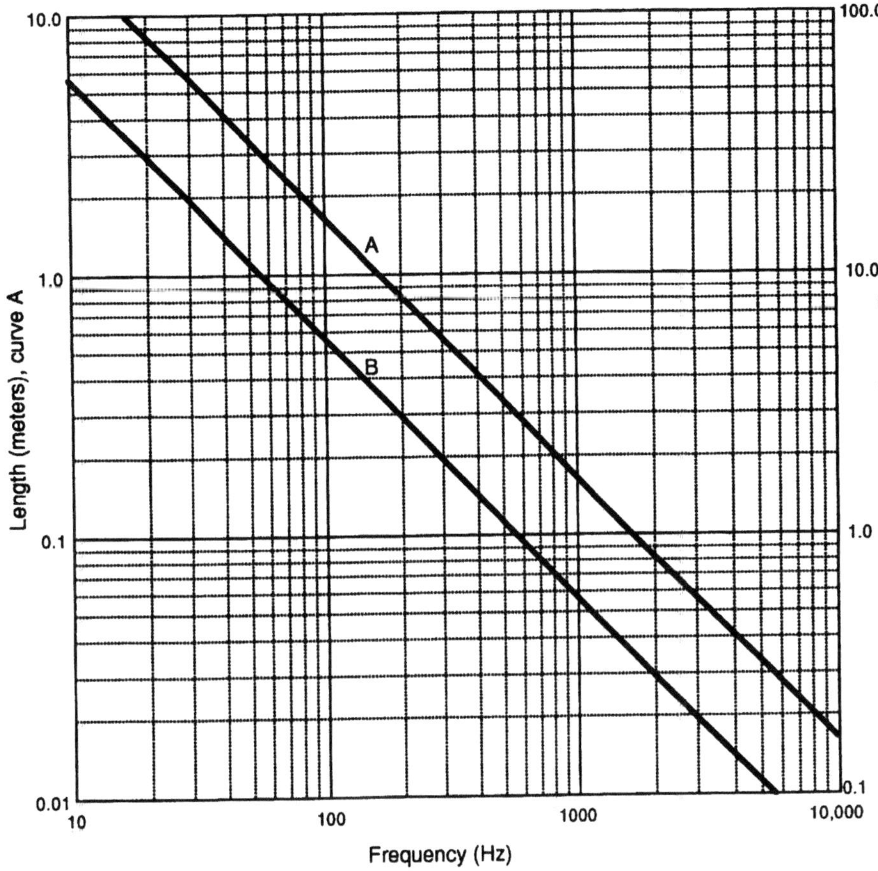

Figure 11. Resonance Frequency for Pipes Open at Both Ends

12. End Correction For Pipes

Assume that the open pipe in the previous example was specified as being of a particular diameter. We now want to determine what degree of end correction must be incorporated to attain the desired frequency. Figure 12 shows the relation between the increase in the speaking length of pipe and its length-to-radius ratio (L/R).

EXAMPLE: Assume a diameter of 0.07 meter for the pipe in the previous example. We now want to find the actual speaking length of the pipe. First, take the ratio of pipe length divided by its radius. The radius is 0.035 meter, and the ratio of length to radius is:

Ratio = 0.7/.035 = 20

Locate 20 along the abscissa and move up to the bold line marked A (an open pipe is effectively open at *both* ends). The ordinate is intersected at a point equal to 6%. This indicates that the speaking length of the pipe will be:

Speaking length = (0.7)(1.06) = 0.742 meters

Will this pipe resonate at 250 Hz? Not quite; it will be a little low in pitch:

Actual pitch = (339)(0.5)/0.742 = 228.4 Hz

Therefore, the pipe length will have to be recalculated in order to determine the correct length.

Reference: Backus (1969)

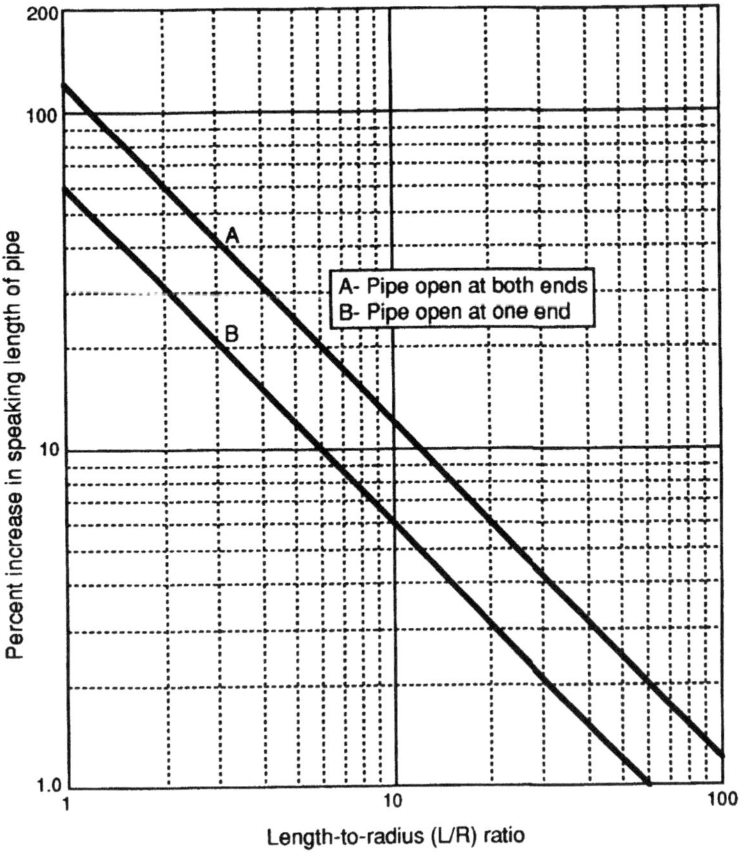

Figure 12. End Correction for Pipes

13. Resonance Frequency for Pipes Open at One End

In effect, a stopped pipe is open at the mouth end and stopped at the other. Its resonance frequency has a wavelength that is nominally one-fourth the length of the pipe. Figure 13 shows the relation between the length of the pipe and its resonance frequency.

EXAMPLE: Determine the length of a stopped pipe that has a frequency of 250 Hz.

On Figure 13, locate 250 Hz along the abscissa, and move upward to the bold line marked A. The intersection takes place at about 0.35 meter along the left ordinate.

Assume that this pipe has a diameter of 0.07 meter. Now go back to section 12 and find the actual speaking length of the pipe. With a radius of 0.035 and a length of 0.35 meter, the ratio of length over radius is 10.

Referring to Figure 12, locate the value 10 along the abscissa and move upward to the bold line marked B. The intersection takes place at a value of 0.6, indicating that this pipe will have an actual speaking length of:

Speaking length = (0.35)(1.10) = 0.385 meter

Actual pitch = (339)(0.25)/.385 = 220 Hz

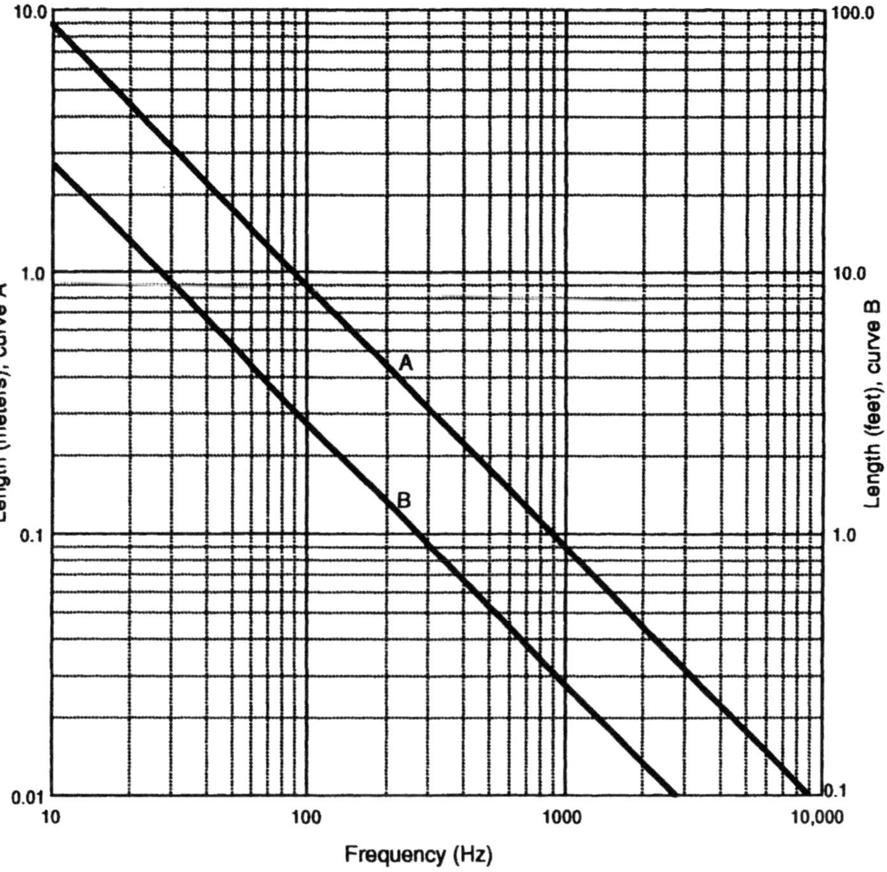

Figure 13. Resonance Frequency for Pipes Open at One End

14. Diffraction of Sound by a Cylinder, a Cube, and a Sphere

The data presented in Figure 14 show the effect on three objects by the diffraction of a plane wave impinging on the front surface. The values of relative pressure level are given for a number of angles of incidence. The plots show the level of the pressure that exists at the front surface. Thus, they give a good indication of the nominal response of a microphone of the same shape.

EXAMPLE: Assume that a microphone is of cylindrical shape and has a diameter of .017 meter. Determine the relative response at 10 kHz for a plane wave incident on the front at 0°.

First, calculate the diameter-to-wavelength ratio (refer to section 2 for the wavelength):

Ratio = (0.017)/.034 = 0.5

We can now go directly to the graph and read, at an angle of 0°, that the response will be about 8 dB greater than at lower frequencies.

Most microphones are small enough that their diffraction effects normally takes place in the range of diameter/wavelength equal to unity or less.

References: Olson (1957)
Muller, Dunn, and Black (1938)

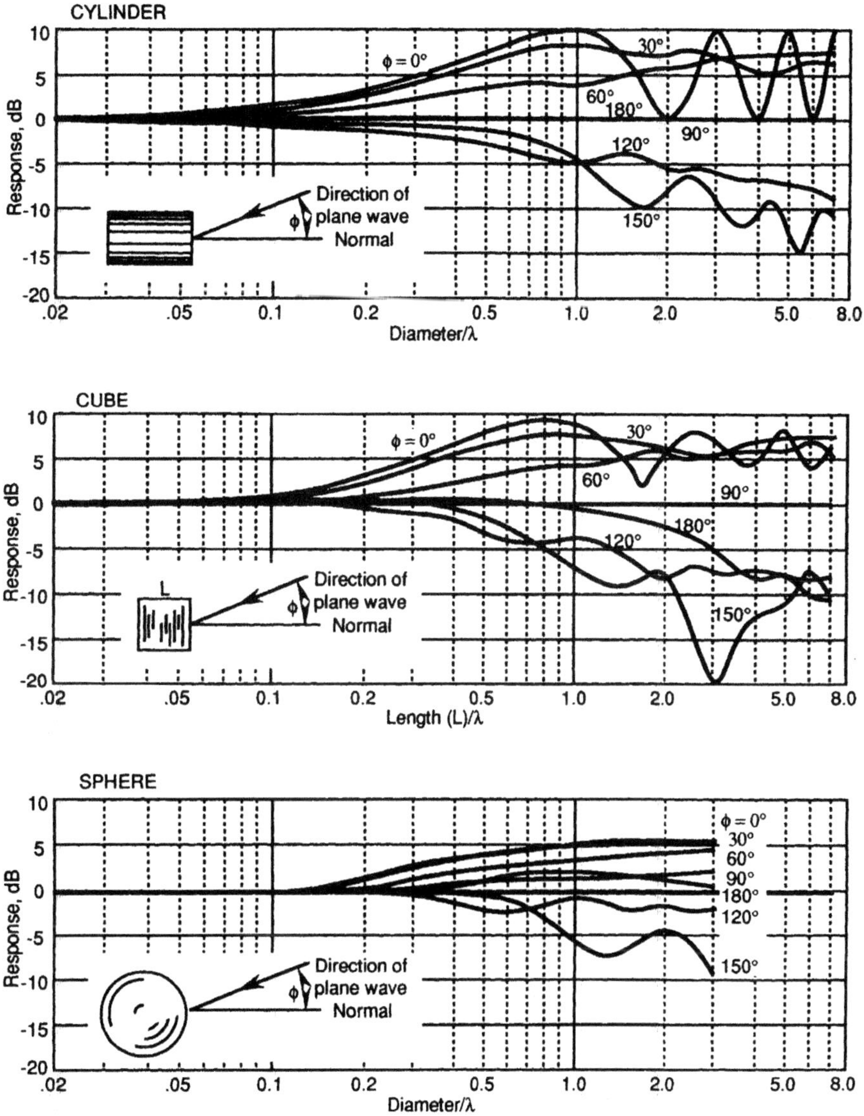

Figure 14. Diffraction of Sound by a Cylinder, a Cube, and a Sphere

15. Response Curves Showing Diffraction by 10 Objects of Different Shape

The data given in Figure 15 show the measured on-axis response of a small loudspeaker placed, as shown, on several objects of the general dimensions indicated. It should be clear that shapes that avoid sharp transitions give a smooth response.

In particular, the sphere (upper left) and the truncated pyramid arrangement (lower right) have a nearly identical response. It is no coincidence that these shapes have been favored by loudspeaker designers for many years.

Reference: Olson (1957)

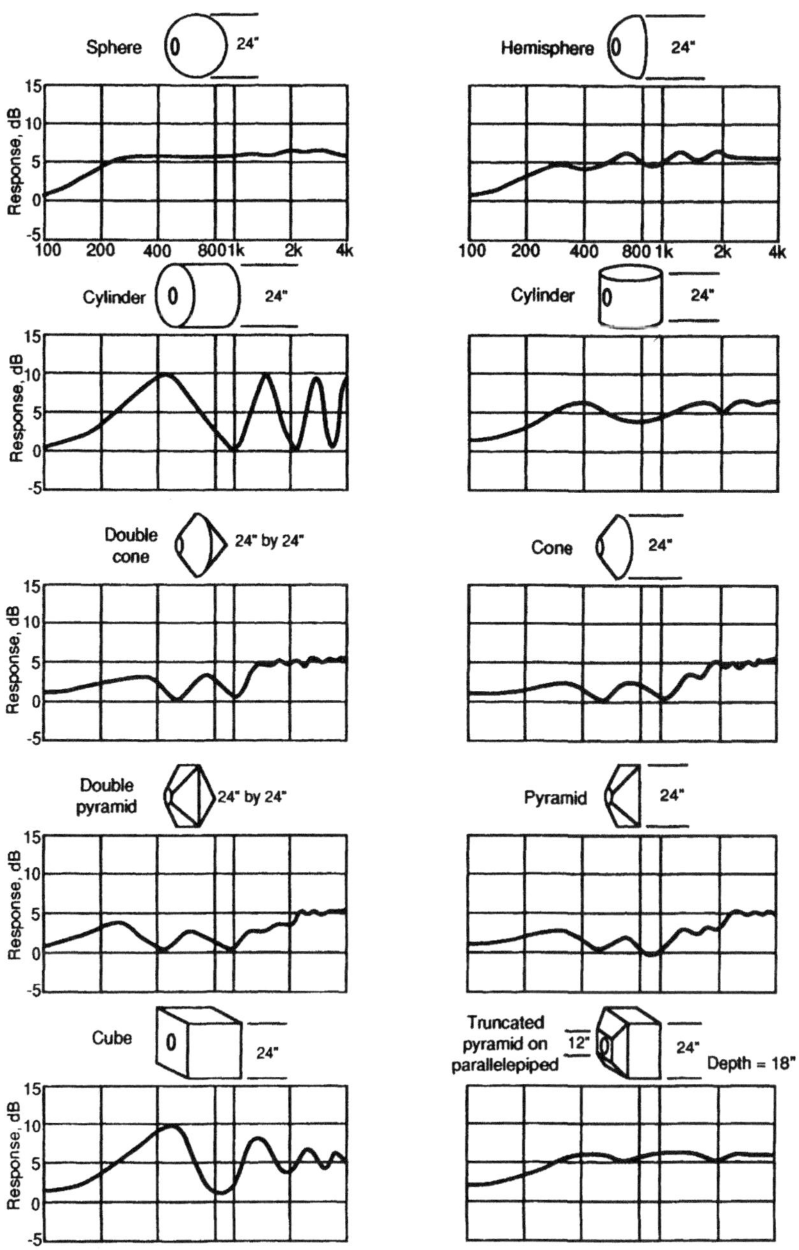

Figure 15. Response Curves Showing Diffraction by 10 Objects of Different Shape

16. Fresnel Diffraction over Sound Barriers

Freeways are often isolated from nearby residential areas by barriers in order to cut down on noise from vehicles. The barrier height is determined from the desired amount of noise loss. In Figure 16A, S represents a line source, such as a busy freeway, and the barrier is actually a long wall viewed along its end. R represents a target residence where a certain amount of sound attenuation is desired, as compared to the case where there is no barrier.

The Fresnel number, N, can be determined directly from the equation. The various distances are indicated, and λ represents the wavelength of sound.

S can also represent a point source close to the ground. In this case, barrier attenuation can be determined from the indicated curve.

EXAMPLE: Determine the Fresnel number (N) for an incoherent line source for the following dimensions at a frequency of 150 Hz:

$d = 40$ meters
$d_1 = 15$ meters
$d_2 = 35$ meters
$\lambda = 2.3$ meters

$$N = (2/\lambda)(d_1 + d_2 - d) = 0.87(15 + 35 - 40) = 17.4$$

By inspection, a Fresnel number of 17.4 corresponds to an attenuation of about 17 dB for the incoherent line source. Determination of the barrier height needed is a simple geometric exercise.

Reference: Harris (1979)

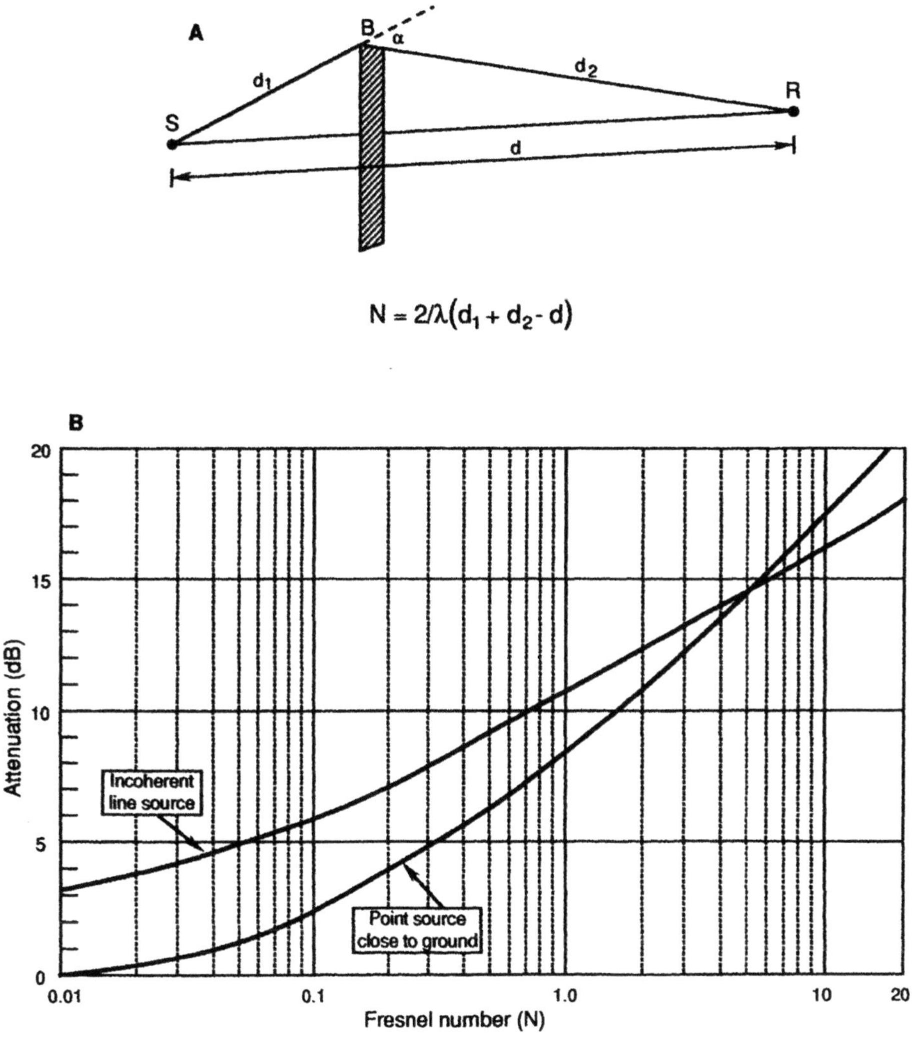

Figure 16. Fresnel Diffraction over Sound Barriers

17. Definition of Critical Distance

Figure 17 is the first of many that will relate direct and reverberant field measurements in large enclosed spaces. The equation that gives the reduction in sound pressure as we move away from a point source is as follows:

$(P)^2 = W_A \rho_0 c(Q/4\pi r^2 + 4/R)$ . . . . (17.1)

We can state this loss in decibels as follows:

$L_p = 10 \log W_A + 10 \log \rho_0 c + 94 + 10 \log$
$(Q/4\pi r^2 + 4/R)$ . . . . (17.2)

where: W_A = power emitted by source (acoustic watts)
$\rho_0 c$ = acoustic impedance of air (407 acoustic ohms)
r = distance from source (meters)
Q = directivity factor (dimensionless) in the direction of observation
R = room constant, $S\bar{\alpha}/(1 - \bar{\alpha})$, where S is the room boundary area in square meters and $\bar{\alpha}$ is the average absorption coefficient of the room (dimensionless)

The first term in the parentheses determines the direct field component of sound level; the second term determines the reverberant field component. We can simplify the equation if we look at the two components separately:

$L_{reverberant} = 126 + 10 \log W_A - 10 \log R$ 17.3

$L_{direct} = 109 + 10 \log W_A + 10 \log Q - 20 \log r$ 17.4

Critical distance, D_C, is defined as the distance from the source at which both reverberant and direct field components are equal. This can be derived simply by equating the two term in the parentheses in equation 17.1 and solving for r:

$Q/4\pi r^2 = 4/R$

$r^2 = (R/4)(Q/4\pi) = QR\ (.0199)$

Finally,

$r = D_c = .14\sqrt{QR}$ 17.5

Reference: Beranek (1954)

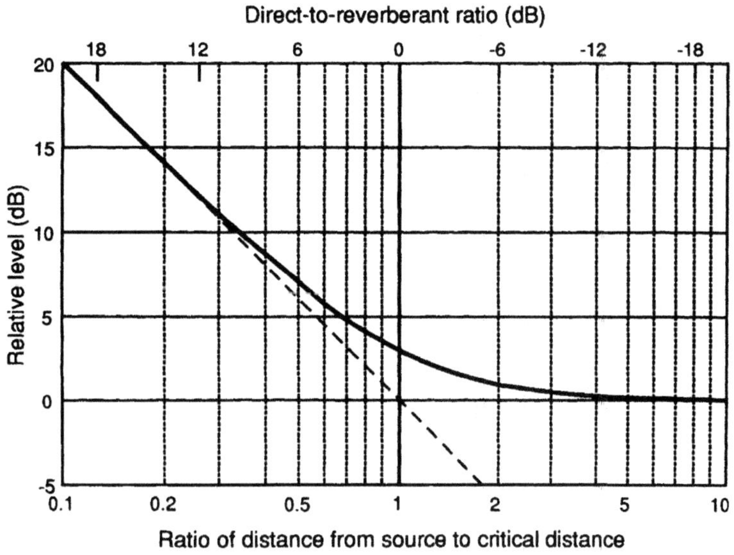

Figure 17. Definition of Critical Distance

18. Room Constant as a Function of Surface Area and Absorption

Room constant (R) identifies the effective absorption in a space in terms of square meters of totally absorptive surface. Its definition was given in section 17.

Figure 18 provides a convenient way of arriving directly at the room constant where both surface area and average absorption coefficient are known.

EXAMPLE: In a room that has surface area of 50,000 square meters and an average absorption coefficient of 0.16, determine the room constant.

Locate 50,000 on the abscissa and move upward to intersect with the bold line labeled 0.16. Then move directly to the ordinate at the left of the graph and read the value for the room constant. In this case, it is 10,000 square meters.

The graph can also be used for calculations with *English* dimensions. However, remember that values of R should always carry the designation or square meters of square feet to make it clear which system you are using.

Reference: Beranek (1954)

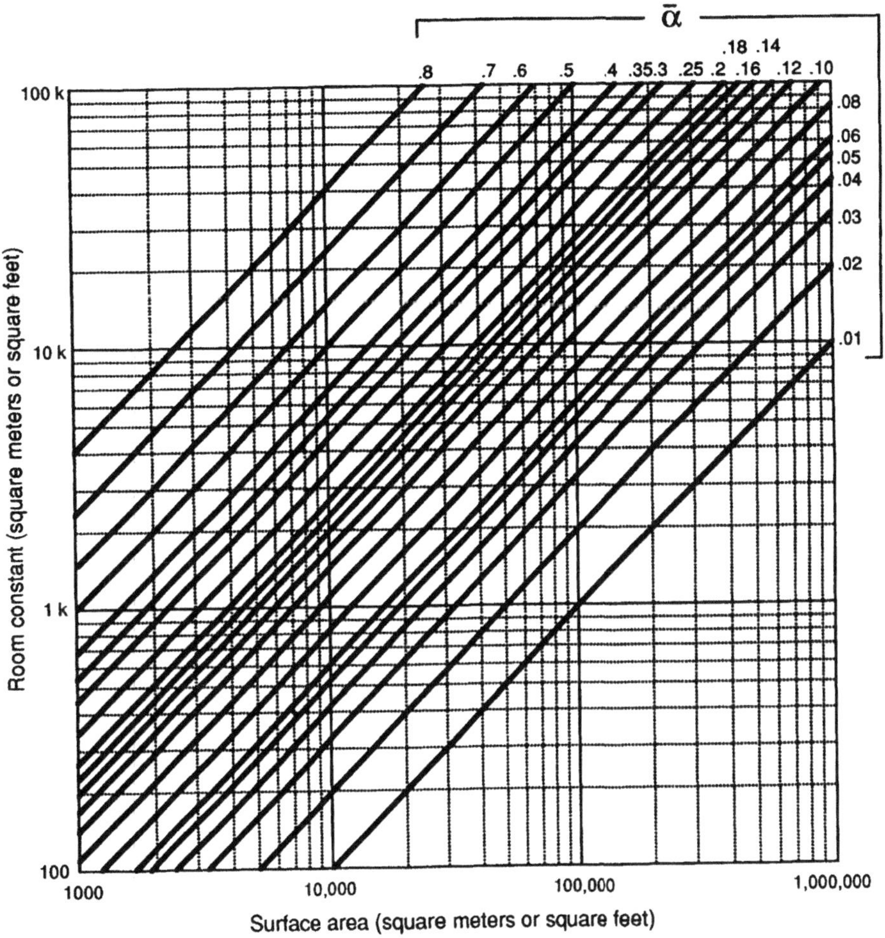

Figure 18. Room Constant as a Function of Surface Area and Absorption

19. Relation Between $\overline{\alpha}$ and $-\ln(1-\overline{\alpha})$ in Reverberation Time Calculations

The Sabine reverberation time equation is given by:

$$T_{60} = (.16V)/S\overline{\alpha} \quad (19.1)$$

The Norris-Eyring equation is given by:

$$T_{60} = (.16V)/-S\ln(1-\overline{\alpha}) \quad (19.2)$$

In spaces that have relatively low absorption (and long reverberation times), both equations give essentially the same results. However, in spaces that have relatively high absorption, the equations will yield results that diverge from each other.

The nomograph in Figure 19 compares values of $\overline{\alpha}$ and $-\ln(1-\overline{\alpha})$; for values of $\overline{\alpha}$ up to about .19, the two track fairly well. At higher values of $\overline{\alpha}$, the divergence is considerable, approaching 25% at $\overline{\alpha} = .40$.

In theoretical terms, the Norris-Eyring equation is more accurate than the Sabine equation, since it will give a value of reverberation time that approaches zero in the case of an average absorption coefficient that approaches unity (total absorption). By comparison, the Sabine equation will give a finite value of reverberation time when the average absorption coefficient approaches unity.

Nevertheless, many consultants prefer to use the Sabine equation.

References: Beranek (1954)
Knudsen and Harris (1950)

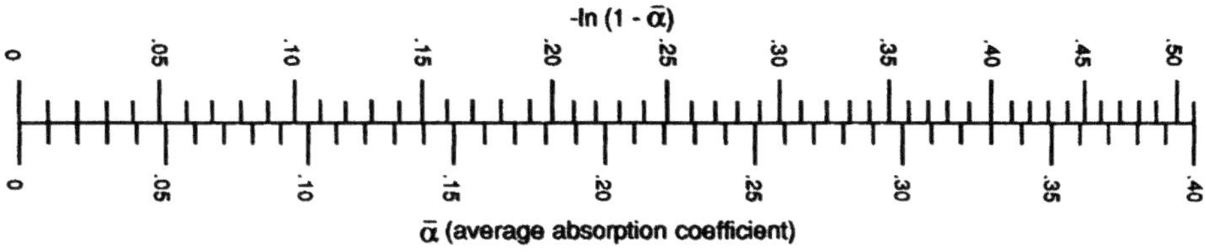

Figure 19. Relation between $\bar{\alpha}$ and –ln (1–α) in Reverberation Time Calculations

20. Reverberant Level as a Function of Room Constant and Acoustical Power

Figure 20 presents solutions to the equation:

$$L_{reverberant} = 126 + 10 \log W_A - 10 \log R \quad \ldots\ldots\ldots\ldots\ldots (20.1)$$

In English units, the equation is:

$$L_{reverberant} = 136 + 10 \log W_A - 10 \log R \quad \ldots\ldots\ldots\ldots\ldots (20.2)$$

EXAMPLE: In a space with a room constant of 5000 square meters and a sound source of .4 acoustical watts, find the corresponding reverberant level in the space.

Locate 5000 on the abscissa and move upward to the bold line that corresponds to acoustical power of .4 watts. Then move across to the ordinate and read the level of 85 dB L_p.

The graph can also be used to determine L_p for acoustical powers outside the range given there.

EXAMPLE: Find the reverberant level produced by a source that delivers 8 acoustical watts into a room with room constant of 5000 square meters.

First, divide 8 watts by 10 to give the value of .8 watts. Proceed as before to arrive at an L_p value of 88 dB. Then, add 10 dB to that number to arrive at the final value of 98 dB.

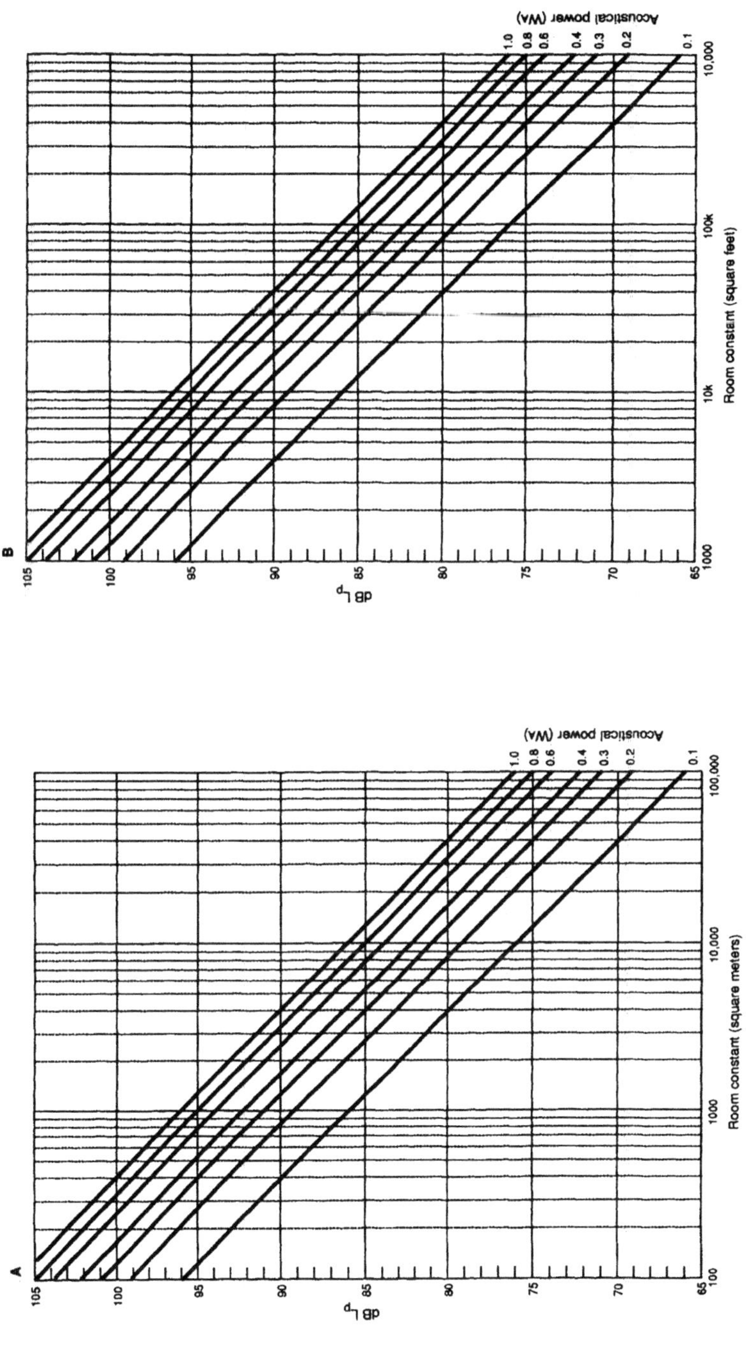

Figure 20. Reverberant Level as a Function of Room Constant and Acoustical Power

21. Mean Free Path (MFP), Room Volume, and Surface Area

MFP, as shown in Figure 21, is a statistical quantity that indicates the average path length taken by sound as it is reflected from one boundary to another. The equation for MFP is:

$$MFP = 4V/S \quad (21.1)$$

where V is the room volume and S is the room boundary area. The equation is applicable to any units of measure.

EXAMPLE: For a room with a volume of 10,000 cubic meters and a surface area of 2000 square meters, find the MFP.

Locate the volume along the ordinate and the area along the abscissa, then find their intersection on the graph. This takes place at an MFP of approximately 20 meters.

References: Beranek (1954)
Cremer and Müller (1978)
Kuttruff (1973)

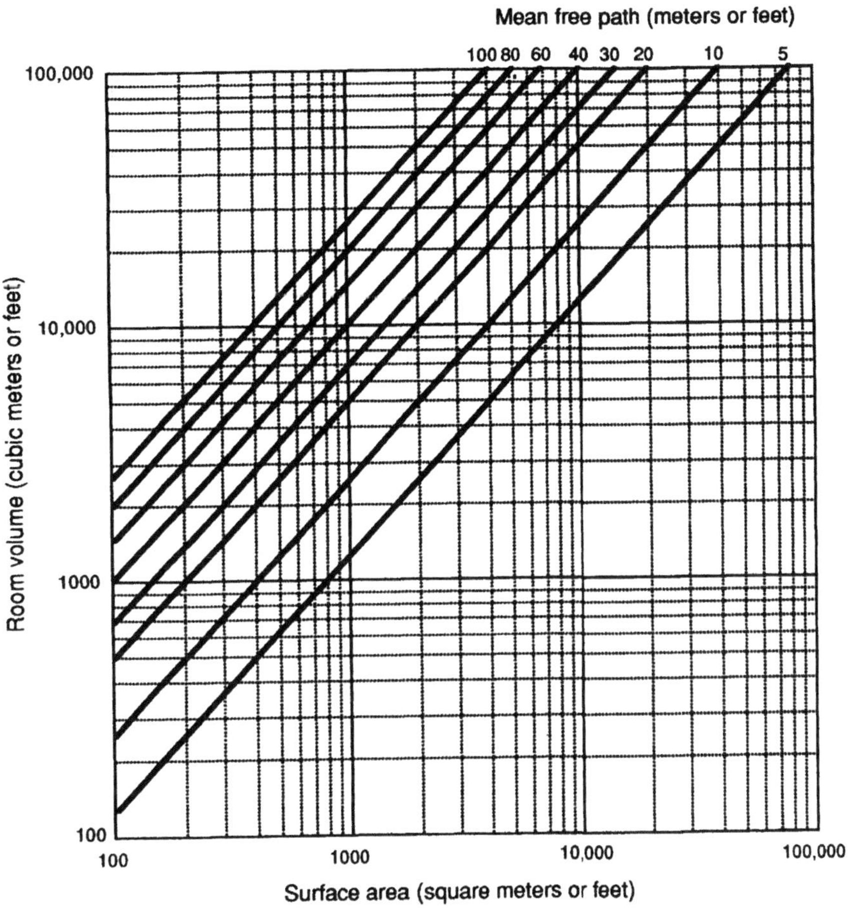

Figure 21. Mean Free Path (MFP), Room Volume, and Surface Area

22. SOUND ATTENUATION OVER DISTANCE IN SEMIREVERBERANT SPACES

In many spaces, particularly those with relatively low ceilings and high absorption, there is no measurable statistical reverberant field, even though equation 17.3 indicates that there should be one. There will be the usual inverse square loss out to critical distance, but beyond critical distance, the observed attenuation will be between the statistical reverberant level and the inverse square level.

The basic attenuation pattern is shown in Figure 22; the actual slope beyond critical distance will depend on the specific characteristics of the space.

Peutz suggests the following equation for estimating the attenuation slope beyond critical distance in rooms with very low reverberation times:

$$\Delta = 0.4 \frac{\sqrt[6]{V}}{T_{60}} \quad \ldots\ldots\ldots\ldots\ldots\ldots\ldots\ldots\ldots\ldots\ldots \quad (22.1)$$

where Δ indicates the falloff in level per doubling of distance, V is the room volume in cubic meters, and T_{60} is the reverberation time in seconds.

EXAMPLE: For a room with a volume of 10,000 cubic meters and a calculated reverberation time of .4 seconds, determine the approximate attenuation beyond the calculated critical distance.

The calculation is performed as follows:

$\Delta = 0.4(10{,}000)^{1/6}/0.4$

$\Delta = 0.4(4.7)/0.4 = 4.7$ dB per doubling of distance

Schultz notes that a general falloff of 3 dB per doubling of distance beyond critical distance will often be observed in many semireverberant spaces.

In practice, it is difficult to estimate behavior in semireverberant spaces with anything better than a very broad stroke. The information given here should only be taken as indicative of general trends.

References: Peutz (1973)
Schultz (1984)

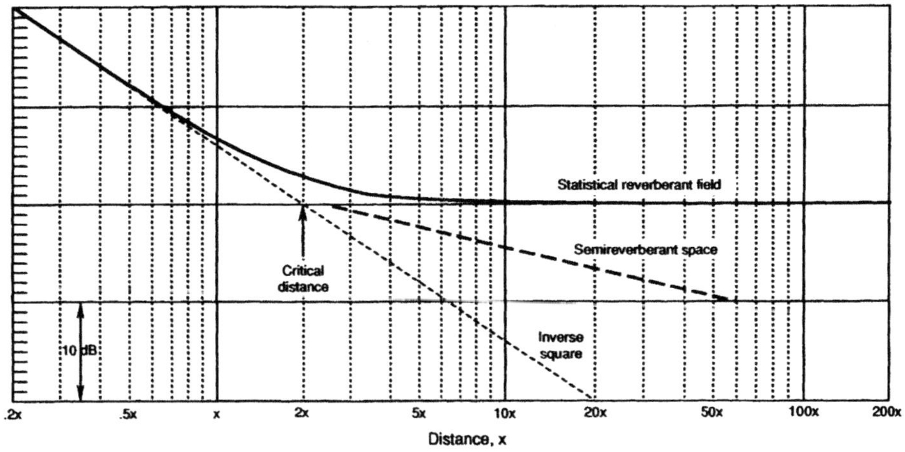

Figure 22. Sound Attenuation over Distance in Semireverberant Spaces

23. Critical Distance as a Function of Room Constant and Directivity Factor

Figure 23 presents solutions for equation 17.5

$$D_C = .14 \sqrt{QR}$$

The ordinate at the right of the graph is in directivity factor, which is normally designated by Q. Directivity index (DI) is given at the left ordinate. DI is defined as follows:

$$DI = 10 \log Q$$

and it is given in decibels.

EXAMPLE: For a Q of 10 and a room constant of 2000 square meters, determine the critical distance.

Locate the value of 2000 square meters along the abscissa and the directivity factor of 10 along the right ordinate. Determine their intersection point, which corresponds to the bold diagonal line labeled 19.8 meters D_C.

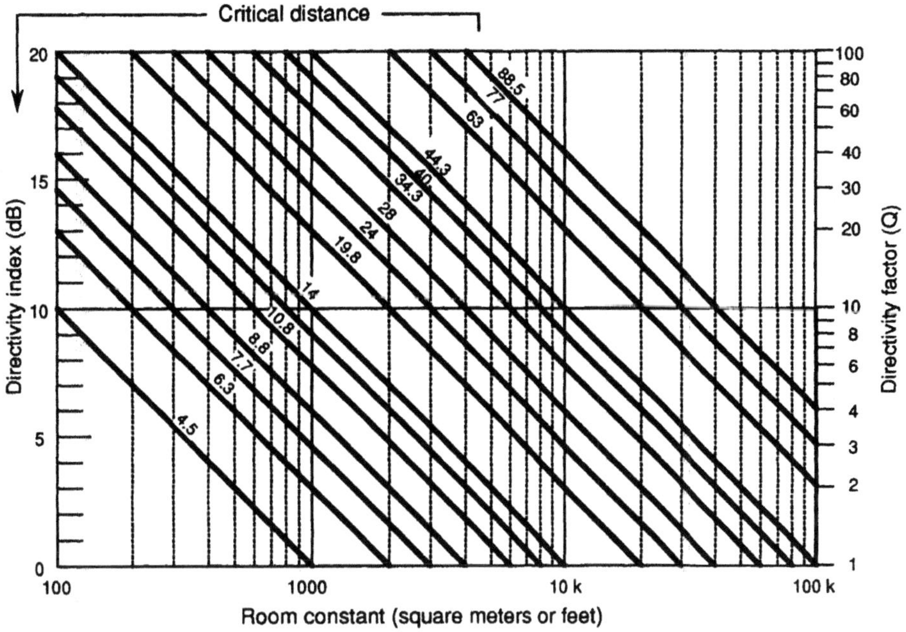

Figure 23. Critical Distance as a Function of Room Constant and Directivity Factor

24. Acoustical Power Required to Produce a Level of 94 dB L_P as a Function of Room Volume and Reveberation Time

Figure 24 represents a set of solutions to the following metric equation set:

$$L_{reverberant} = 126 + 10 \log W_A - 10 \log R \quad (24.1)$$

$$R \cong S\bar{\alpha} = (.16)V/T_{60}. \quad (24.2)$$

in which the reverberant level has been fixed at 94 dB L_p. The curves on the graph represent various reverberation times. The data are fairly accurate in rooms that are particularly "live"—that is, in rooms with an average absorption coefficient in the range of 0.2 or less.

EXAMPLE: For a room with a volume of 10,000 cubic meters and a reverberation time of 5 seconds, find the acoustical power required to produce a reverberant L_p of 94 dB.

Locate 10,000 along the abscissa and move upward to the bold line marked 5 seconds. Then move to the ordinate and read 0.2 watts.

For a loudspeaker with a nominal efficiency of 2%, an input power of 10 watts would be required to produce the desired level.

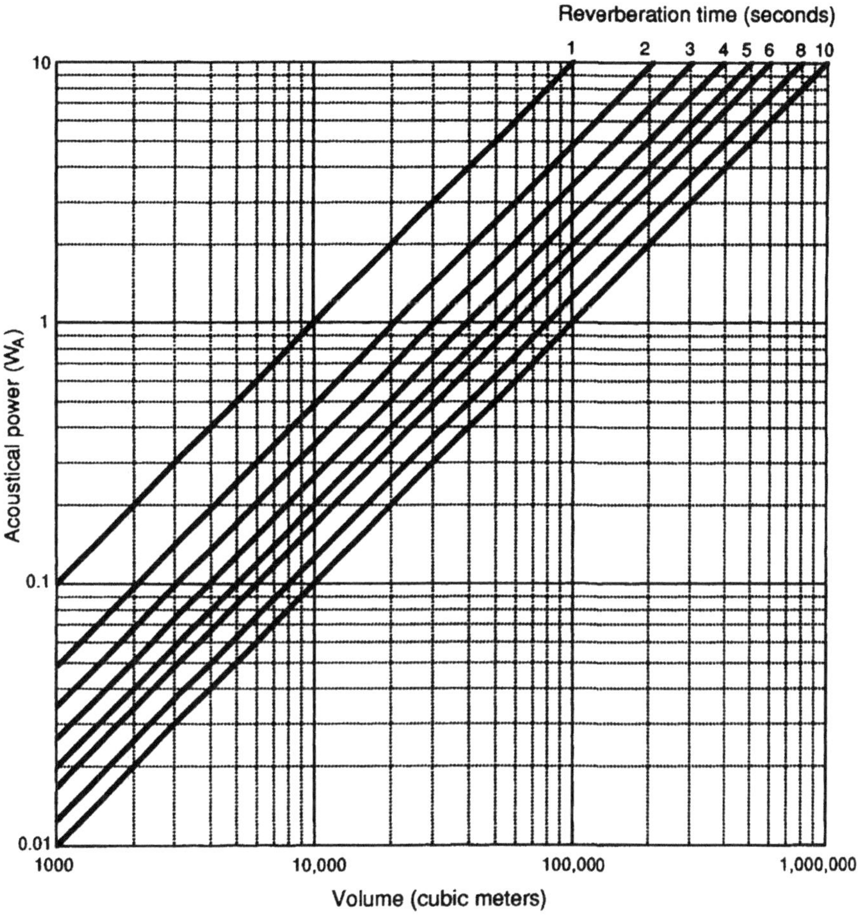

Figure 24. Acoustical Power Required to Produce a Level of 94 dB L_p as a Function of Room Volume and Reverberation Time

25. Sound Pressure Level Produced by 1 Acoustic Watt as a Function of Room Constant and Distance from Source

Figure 25 represents a set of solutions to equation 17.2 in which W_A has been fixed at 1 watt and Q set to unity. The curves on the graph represent several values of room constant.

EXAMPLE: Find the attenuation between distances of 2 and 20 meters from a point source emitting 1 acoustic watt in a room with a room constant of 5000 square meters.

Locate the value of 2 meters along the abscissa and move upward to the curve marked R = 5000. Read the value of approximately 103 dB L_p. Then locate the value of 20 meters along the abscissa and move upward to the curve marked R = 5000. Read the value of 89.5 dB L_p. Taking the difference gives a level difference of 13.5 dB.

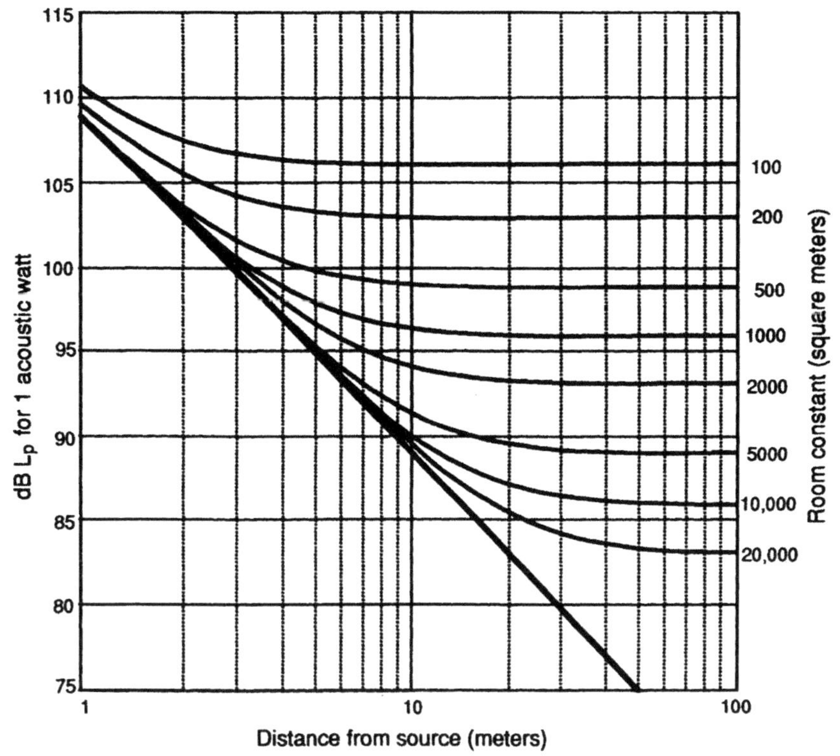

Figure 25. Sound Pressure Level Produced by 1 Acoustic Watt as a Function of Room Constant and Distance from Source

26. Estimation of Total Absorption When Room Volume and Reverberation Time Are Known

The graphs in Figure 26 represent a set of solutions of the Sabine reverberation time equation for both metric and English units.

EXAMPLE: Find the total absorption ($S\bar{\alpha}$ units) in a room with a volume of 5000 cubic meters and a reverberation time of 3 seconds.

Locate 5000 along the abscissa and 3 seconds along the ordinate. Determine their intersection slightly to the left of the bold line marked $S\bar{\alpha} = 300$.

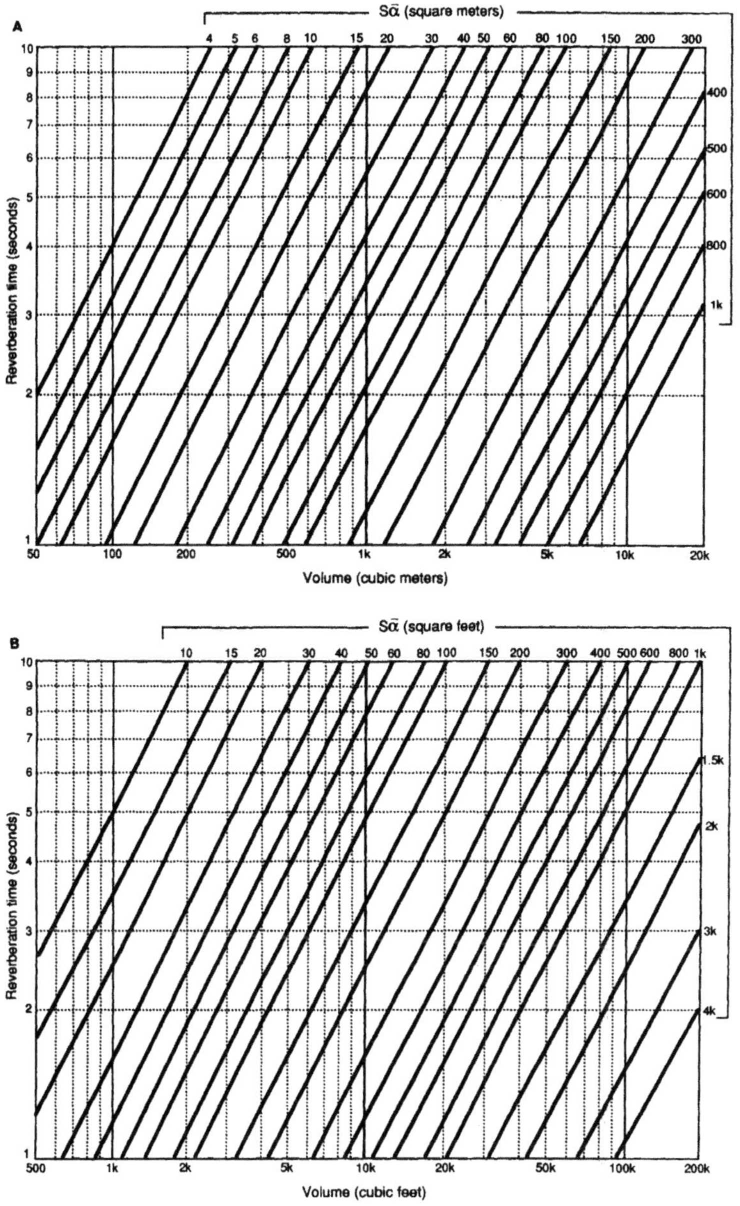

Figure 26. Estimation of Total Absorption When Room Volume and Reverberation Time Are Known

27. Estimation of Room Constant When Room Volume and Reverberation Time Are Known

The graphs in Figure 27 represent a set of solutions of equation 19.2, the Norris-Eyring reverberation time equation. It should provide more accurate estimates at shorter reverberation times. The $-\ln(1-\bar{\alpha})$ correction has been incorporated into the curves. Data are given in both metric and English units.

EXAMPLE: Find the total absorption ($S\bar{\alpha}$ units) in a room with a volume of 4000 cubic meters and a reverberation time of 0.6 seconds.

Locate 4000 long the abscissa and 0.6 along the ordinate. Determine their intersection to the left of the curve marked $S\bar{\alpha} = 1000$. A good estimate of the actual value would be 900 square meters of absorption units.

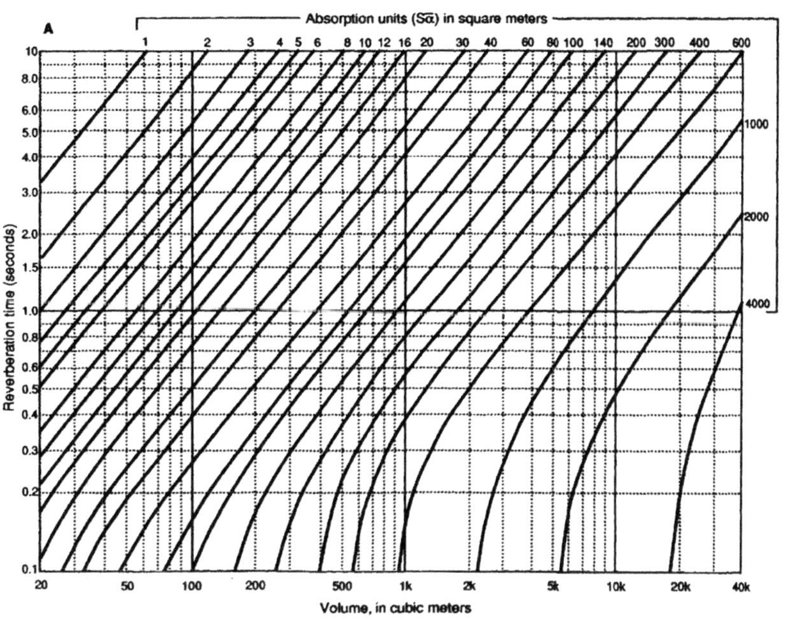

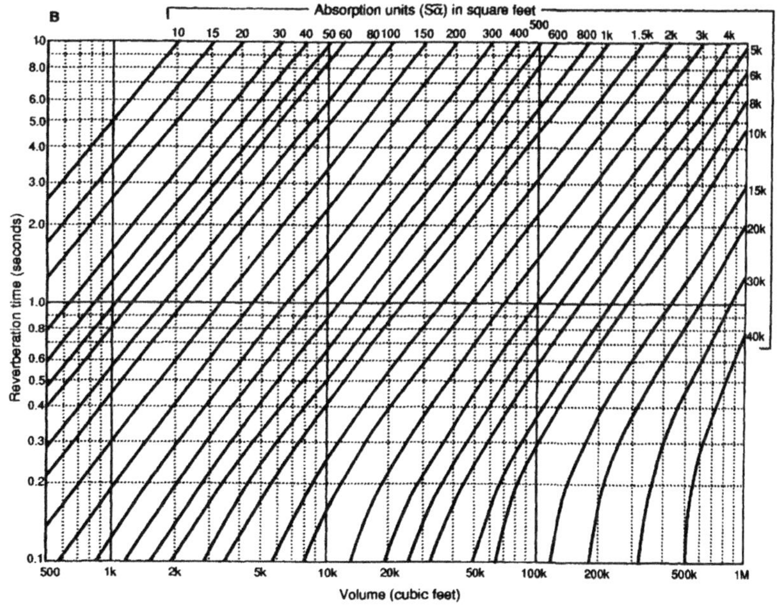

Figure 27. Estimation of Room Constant When Room Volume and Reverberation Time Are Known

28. Estimation of Room Boundary Area When Volume Is Known

In general, it is easier to calculate room volume than to calculate the total boundary area. For rooms that are normally proportioned, such as auditoriums and houses of worship, the approximate relationship between volume and boundary area is given by the following equation:

$$S_T = 6.2\ V^{2/3} \quad \text{(28.1)}$$

where S_T is the boundary area and V is the volume. Any consistent system of units may be used, but for the ranges given in the graph, it will be more convenient to express the data in metric units.

Another useful relationship is given by the following equation:

$$\bar{\alpha} = (.0259\ V^{2/3})/T_{60} \quad \text{(28.2)}$$

The basic relationship given by equation 28.1 is plotted in Figure 28. The small dots indicate sample data points for actual spaces taken from the Acoustical Society of America publication *Acoustics of Worship Spaces.* The small crosses indicate data points for selected Japanese auditoriums taken from the Acoustical Society of America publication *Halls for Music Performance.*

Sepmeyer states that equation 28.1 holds for room dimension ratios that were found to give acceptable modal frequency distribution and that the equation is accurate within a standard error of only 2.5%.

References: Lubman and Wetherill (1985)
Sepmeyer (1988)
Taleske, Wetherill, and Cavanaugh (1982)

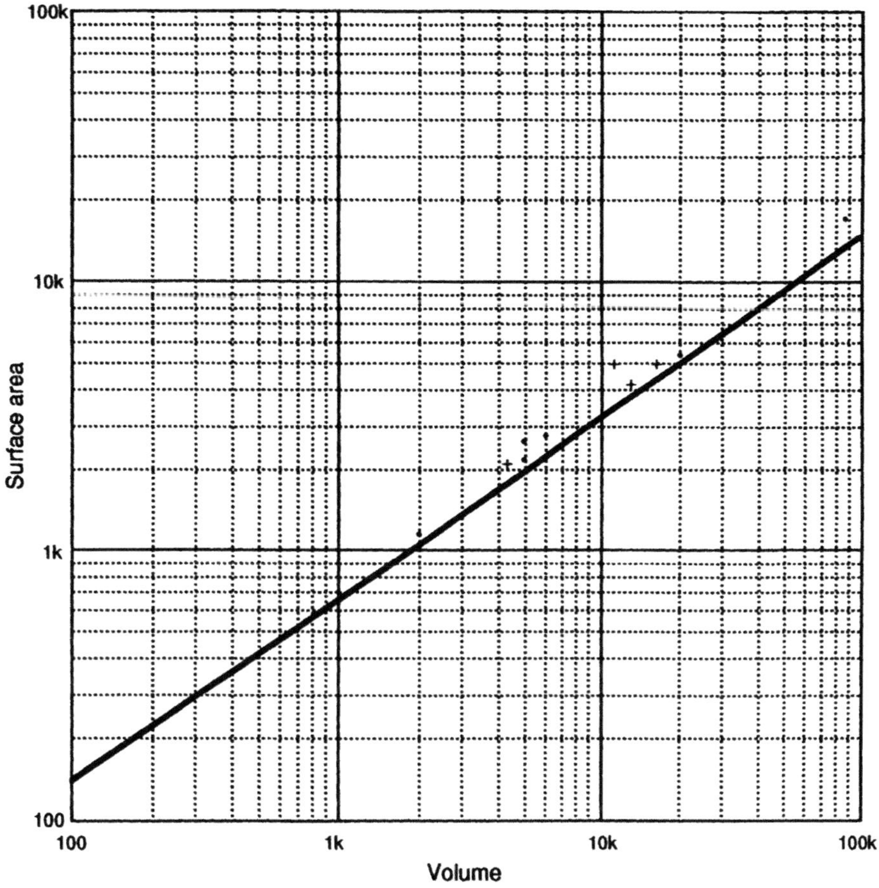

Figure 28. Estimation of Room Boundary Area When Volume Is Known

29. Reverberation Time Ratios with and Without Atmospheric Losses

For reverberation time calculations made using only boundary absorption, the data provided in Figure 29 provide a quick and accurate estimate of the actual reverberation time, considering atmospheric effects at 50% relative humidity.

The curves, which are based on equation 5.1, show the difference in reverberation time with and without the 4mV term in the denominator of the equation.

EXAMPLE: Assume that a room has a calculated reverberation time of 4 seconds, considering only boundary absorption. At a relative humidity of 50%, calculate the reverberation time at 8 kHz.

Locate the value of 4 seconds along the abscissa. Move upward to the bold curve labeled 8 kHz. Then move across to the ordinate and read the value of approximately 1.15 seconds.

Values are given only for 50% relative humidity; this should be sufficient for most estimates of modern rooms with controlled air conditioning.

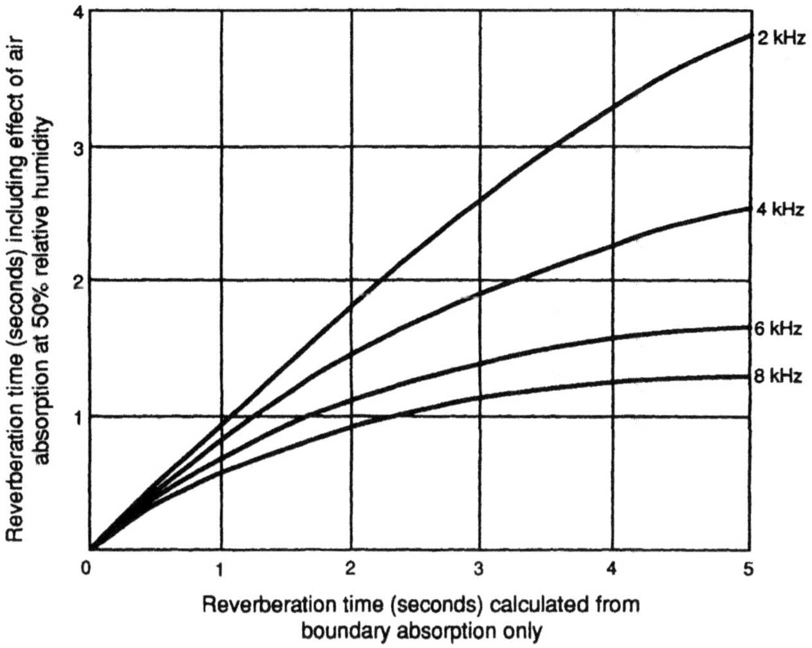

Figure 29. Reverberation Time Ratios with and without Atmospheric Losses

30. Relationship between Directivity Factor and Directivity Index

Figure 30 plots the directivity factor (Q) against the directivity index (DI). As discussed in section 23, Q and DI are useful ways of presenting the same information. Because DI is expressed in decibels, it generally appears in equations where other quantities have been expressed in terms of signal level.

The relevant equations are:

$$DI = 10 \log Q \quad (30.1)$$

$$Q = 10^{DI/10} \quad (30.2)$$

The directivity index along a specific axis of a sound source is defined as the ratio of sound pressure level, measured along that specific axis at a reference distance, compared to the sound pressure level that would be measured at the reference distance if the source were radiating acoustical power uniformly in all directions.

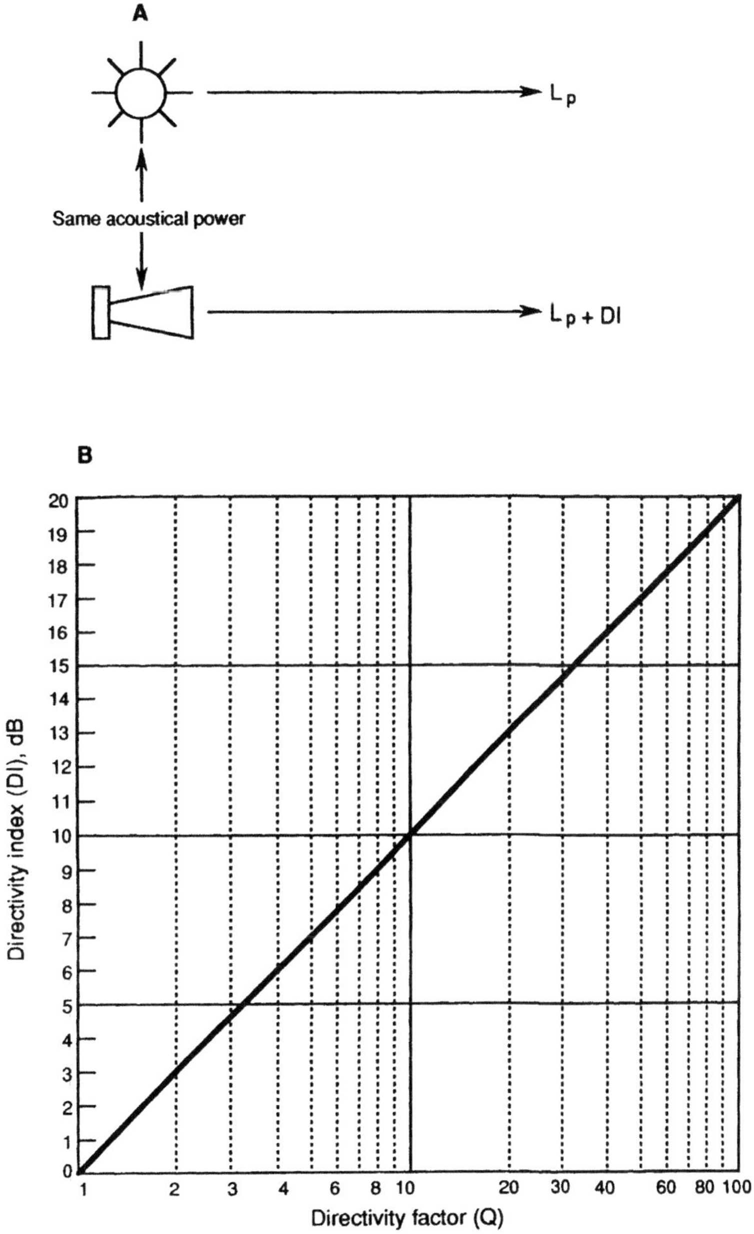

Figure 30. Relationship between Directivity Factor and Directivity Index

31. Wave Number (k) as a Function of Piston Size and Frequency

Wave number (k) relates the size of a radiating piston to the wavelength of the radiated frequency. It is a useful parameter in many graphs showing directivity patterns of loudspeakers and microphones.

Wave number is defined as:

$$k = 2\pi/\lambda \quad (31.1)$$

where λ is the radiated wavelength.

The quantity ka is normally used to relate wave number to a given piston size. Here, a is the piston radius, and ka is thus a dimensionless number.

Since the circumference of a circular piston is $2\pi a$, ka is calculated as follows:

$$ka = \text{circumference}/\text{wavelength} \quad (31.2)$$

For radiating surfaces that are not quite circular, this definition of ka is normally sufficient to give accurate estimates of directional behavior.

Figure 31 shows the relation between frequency and wavelength (meters and centimeters) for sound propagation under normal conditions.

Reference: Beranek (1954)

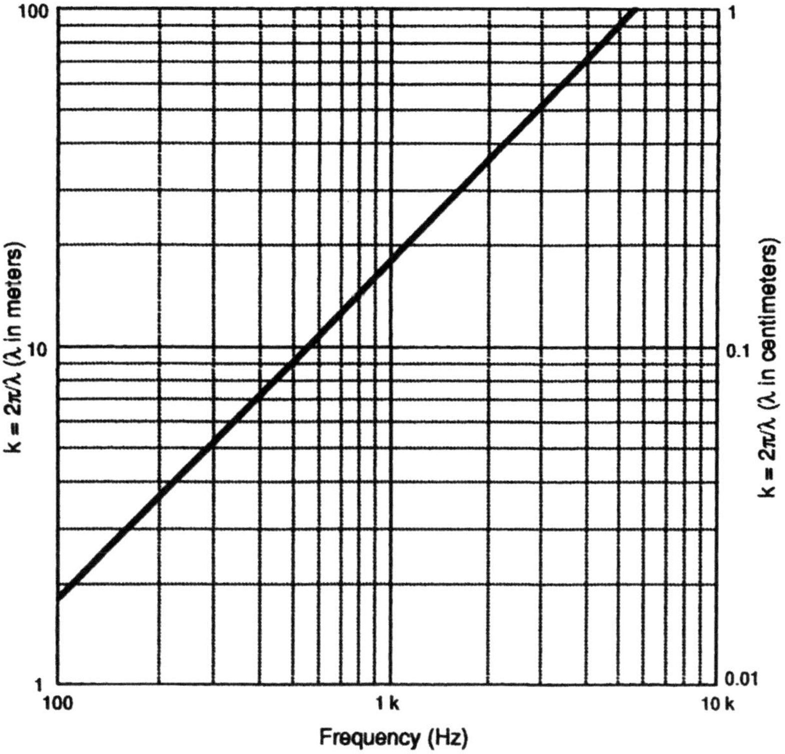

Figure 31. Wave number (k) as a Function of Piston Size and Frequency

32. Polar Response of a piston Mounted in a Large Baffle

The graphs in Figure 32 shows the theoretical polar response of a piston mounted in a large baffle for six values of ka. The directivity index along the zero degree axis is also given.

The directional properties of flush-mounted loudspeakers and microphones are fairly accurately described by the graphs presented in this section.

EXAMPLE: Determine the approximate directional characteristics of a radiating rectangle with dimensions of 0.2 and 0.3 meters mounted in a large wall and operating at a frequency of 500 Hz.

After determining that the circumference of the radiating surface is 1 meter, you can calculate the wavelength of 500 Hz from equation 2.1 as 0.68 meters. Using equation 31.2, ka = 1.48.

By inspecting the graphs, we can see that the response will be somewhere between A and B. Actually, the directional response will be slightly broader along the 0.2 meter axis than along the 0.3 meter axis.

Reference: Beranek (1954)

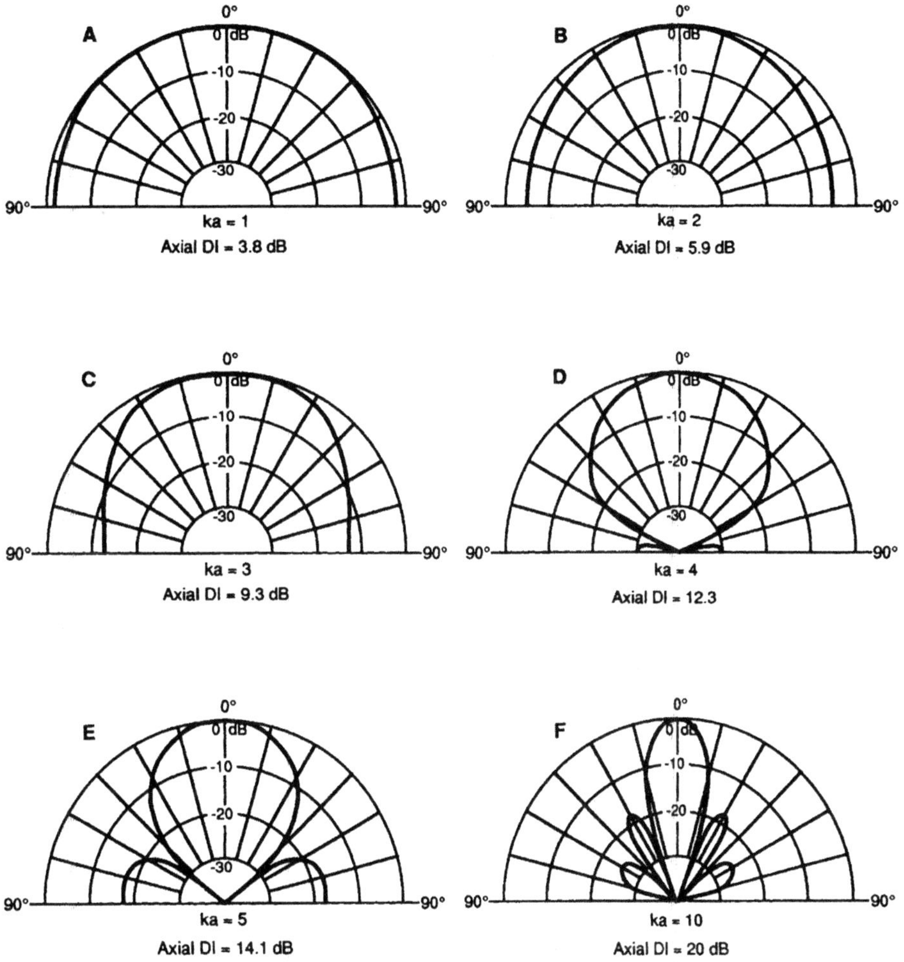

Figure 32. Polar Response of a Piston Mounted in a Large Baffle

33. Polar Response of a Piston Mounted at the End of a Long Tube

The graphs in Figure 33 show the theoretical polar response of a piston mounted in the end of a long tube for six values of ka. The directivity index along the zero degree axis is also given.

These directivity patterns quite accurately describe the performance of a microphone element mounted at the end of a tube.

EXAMPLE: Determine the approximate directional properties of a microphone with a diameter of 1.2 centimeter operating at a frequency of 20 kHz.

The circumference of the microphone is 1.2π centimeters; the wavelength (see section 2) is 0.016 meter, or 1.6 centimeters. Thus,

$$ka = 1.2\pi/1.6 = 2.35$$

The response will then fall about halfway between Figure 33C and Figure 33D, and the DI will be in the range of 3 dB.

Reference: Beranek (1954)

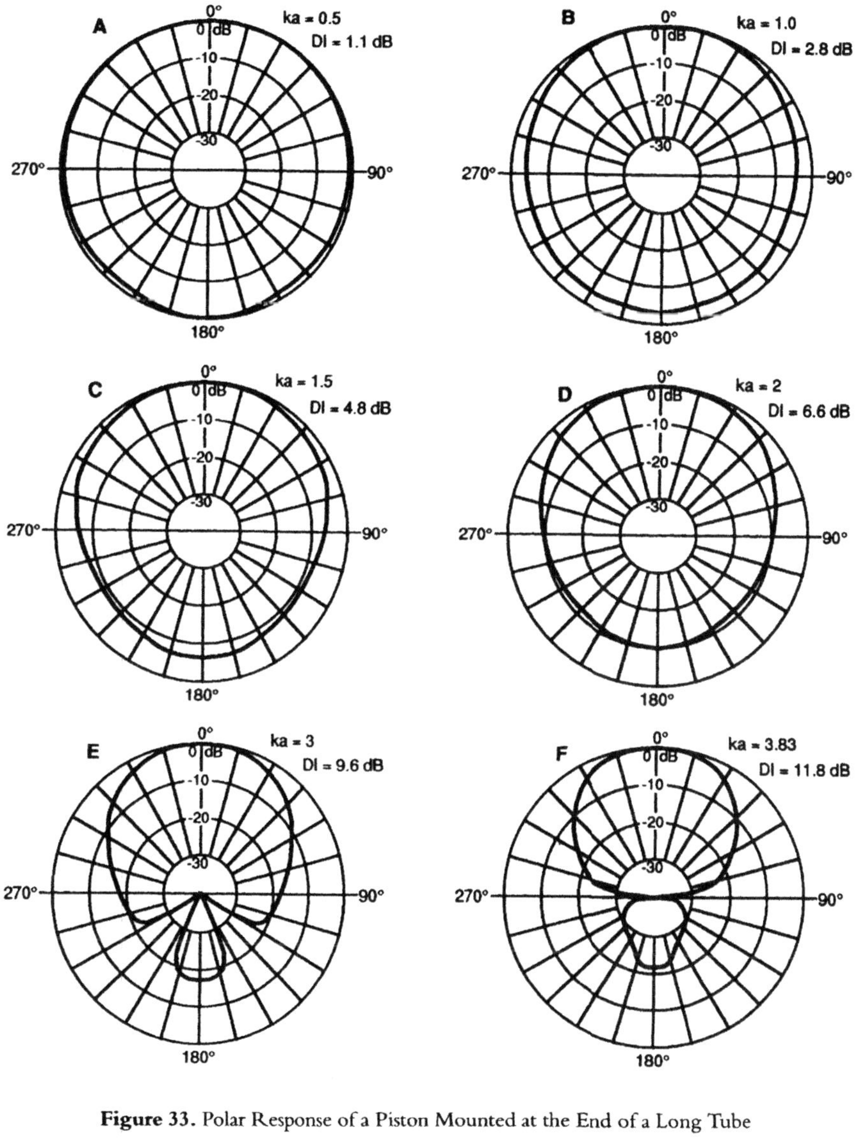

Figure 33. Polar Response of a Piston Mounted at the End of a Long Tube

34. Polar Response of an Unbaffled Piston

An unbaffled piston acts as a dipole; it has a characteristic symmetrical polar pattern at all frequencies that exhibits zero response along the ±90° axes. Furthermore, the front and back response at all frequencies will be in opposite polarity.

Nearly all microphones with figure 8 patterns rely on an unbaffled piston to create the pattern, and most electrostatic loudspeakers are of similar design.

At ka = 1 and lower values, the pressure response is virtually that given by the polar equation:

$$\rho = \text{cosine}\ \theta \qquad (34.1)$$

where the lobes on each side are circular in cross section. However, when plotted in decibels, as at Figure 34A, the response appears to be flattened along the axes, bulging outward to the sides.

EXAMPLE: Determine the approximate directional response of a dipole microphone consisting of an unbaffled piston 3 centimeters in diameter operating at a frequency of 15 kHz.

The circumference of the microphone is 3π, and the wavelength (see section 2) is about 0.02 meter. Thus,

$$ka = 3\pi/2 = 4.7$$

The response then is essentially as shown at Figure 34D.

Reference: Beranek (1954)

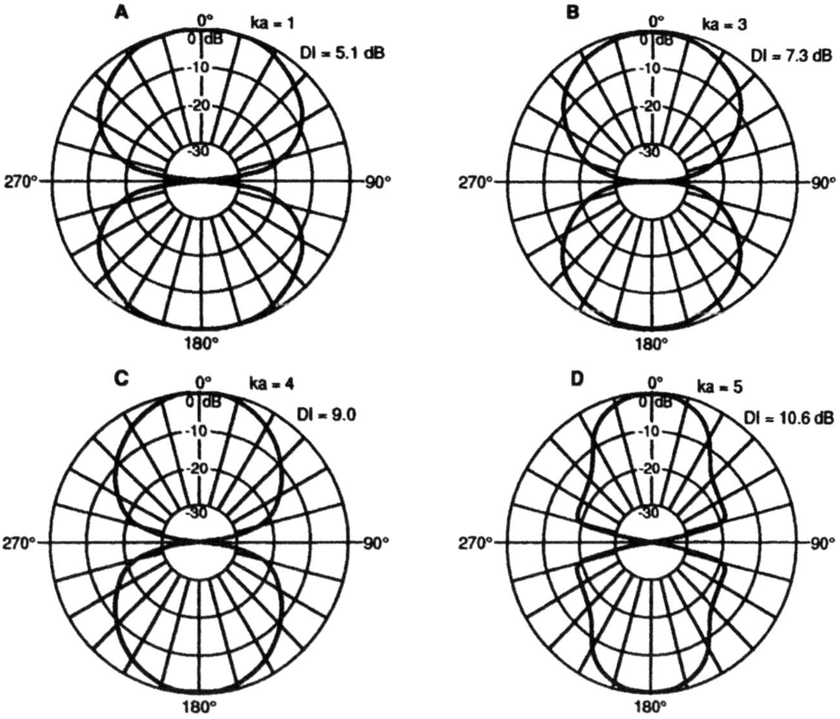

Figure 34. Polar Response of an Unbaffled Piston

35. Off-Axis Response of a Piston in a Large Baffle

The data shown in Figure 35 represent solutions to the following equation:

$$\Gamma(\theta) = \frac{2J_1\left(\frac{\pi d}{\lambda}\sin\alpha\right)}{\frac{\pi d}{\lambda}\sin\alpha} \quad \text{(35.1)}$$

where J_1 represents a Bessel function of the first order, d is the diameter of the piston, λ is the wavelength, and α is the off-axis angle of measurement. Diameter and wavelength must be expressed in the same units.

The range of values covered in the graph are especially useful for determining beamwidth at higher frequencies, where the ratio of diameter to wavelength is large.

EXAMPLE: A loudspeaker 0.2 meters in diameter is radiating a frequency of 2 kHz. Determine the off-axis angle at which the response is 6 dB down.

The radiated wavelength is equal to 339/2000, or 0.17 meter. The ratio d/λ is then 0.2/0.17, or 1.17.

Locate 1.17 along the abscissa and read upward to the intersection with the 6 dB bold curve. The value is approximately 45°.

See section 32 for related data.

References: Olson (1957)
Massa (1942)

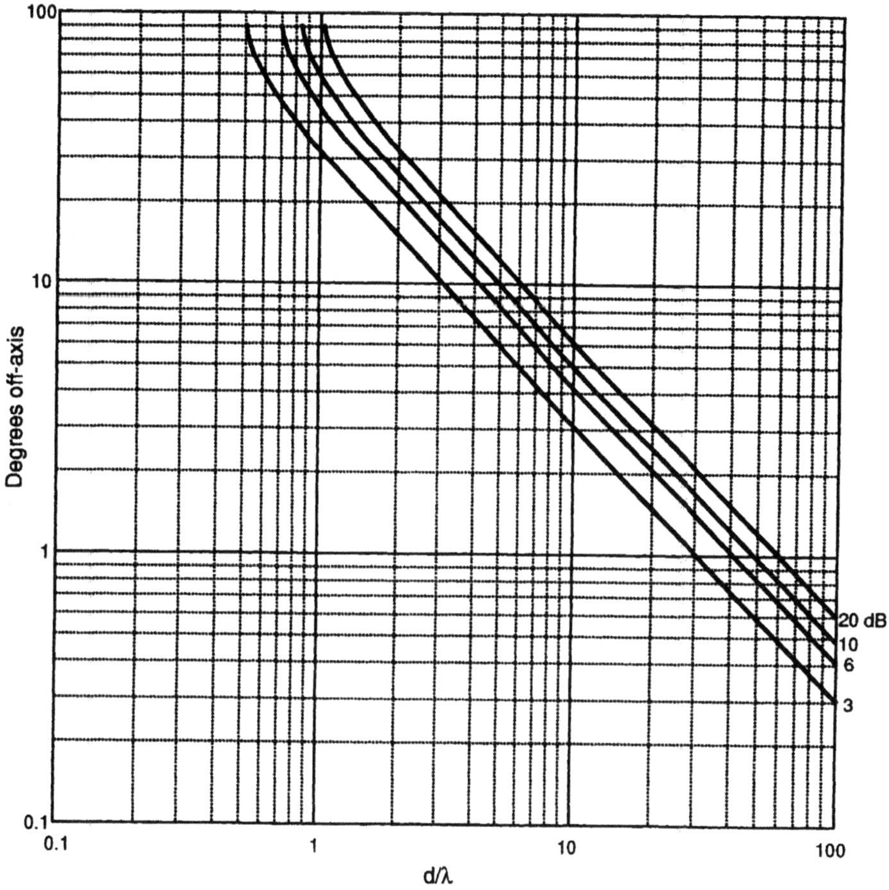

Figure 35. Off-axis Response of a Piston in a Large Baffle

36. Directivity of a Piston in a Large Baffle, at the End of a Long Tube, and in Free Space

Figure 36 gives a summary of the 0° axis directivity index and directivity factor for the three piston mounting conditions discussed in sections 32–34.

Note that, for large values of ka, the directivity of a piston mounted at the end of a long tube or in a large wall converge on the same value. For low values of ka, there is a constant 3-dB difference in the DI, indicative of the fact that wall mounting, with its characteristic half-space radiation, will always have a minimum DI of 3 dB (Q = 2).

By comparison, the unbaffled radiator, regardless of shape, will always have a minimum DI of 4.8 dB (Q = 3) at low values of ka.

Reference: Beranek (1954)

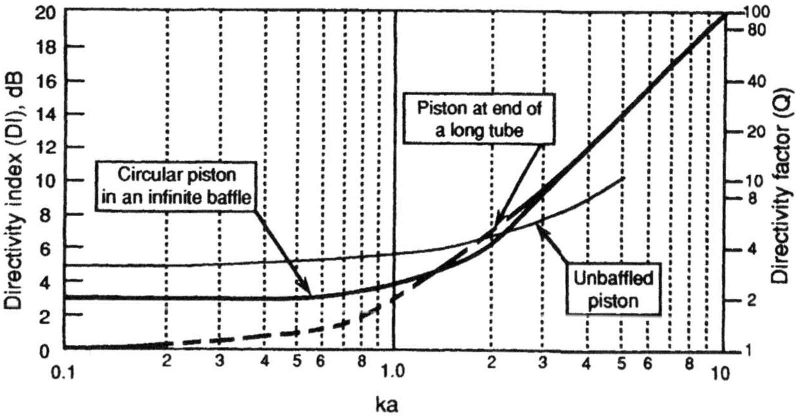

Figure 36. Directivity of a Piston in a Large Baffle, at the End of a Long Tube, and in Free Space

PART TWO

LOUDSPEAKERS

37. Transmission Coefficient versus Frequency for a Piston Mounted in a Large Baffle

Figure 37 is derived from the graph familiar to all students of acoustics showing the radiation resistance for a circular piston mounted in a large baffle. Here we are presenting it as a transmission coefficient with a 6-dB per octave rolloff below f_0.

The normal operating range for a low-frequency cone loudspeaker is bounded at low frequencies by the primary resonance of the moving system and at high frequencies by f_0, as given in the graph. In this frequency region, the loudspeaker cone is mass-controlled, and the falloff in displacement above resonance is compensated for by the rise in transmission coefficient. Over this range, the loudspeaker's power response and on-axis pressure response will be uniform. In most applications, cone loudspeakers may be used well above f_0, taking advantage of the increase in on-axis DI (see section 32).

The following table shows values of f_0 for a number of standard loudspeaker sizes:

Nominal diameter (in)	*Piston diameter (cm)*	*Value of f_0 (Hz)*
18	40.6	417
15	33.0	510
12	25.4	665
10	20.3	835
8	15.2	1110

It can be seen that the time-honored practice of using 15-inch loudspeakers up to 500 Hz takes them just to their upper limit of flat power response. This frequency corresponds to a value of ka of about 1.5, producing directional response that is between Figure 32A and Figure 32B.

References: Augspurger (1971)
Locanthi (1952)

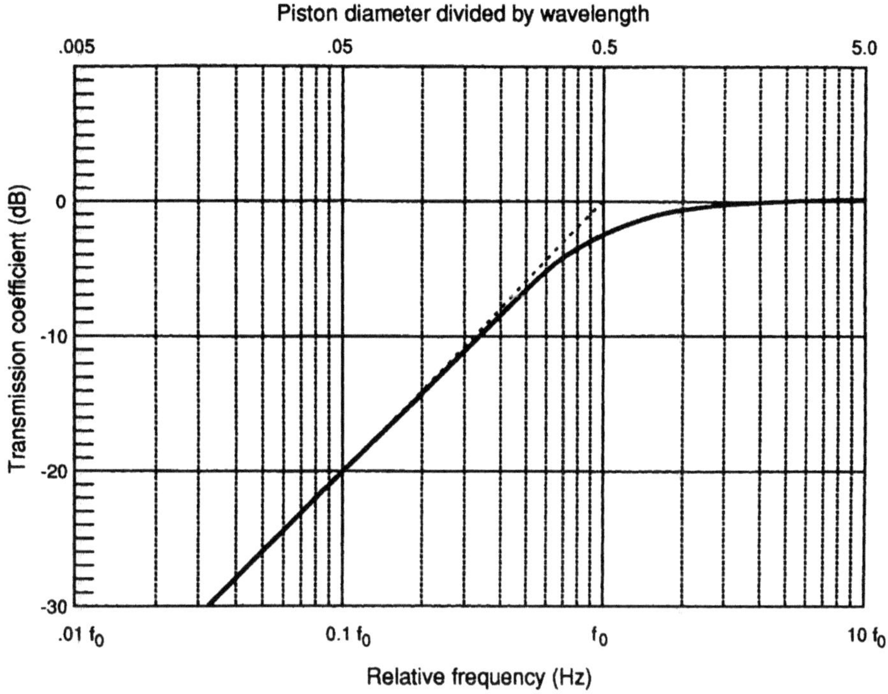

Figure 37. Transmission Coefficient versus Frequency for a Piston Mounted in a Large Baffle

38. Normalized Mutual Coupling for Multiple Pistons

When two identical low-frequency loudspeakers are placed close to one another, there is a tendency for them to couple together and behave as a single loudspeaker with twice the cone area. Under these conditions, we can expect to get four times the power output, or an increase in output of 6 dB. Three decibels come from the simple fact that we have doubled the electrical power input capability, and an additional 3 dB comes from the increase in radiation impedance.

The frequency below which mutual coupling becomes effective is given by the following equation:

$$f_0 = \frac{c}{d}\sqrt{n} \qquad (38.1)$$

where c is the speed of sound (m/sec), d is the nominal spacing between drivers (m), and n is the number of coupled drivers.

Figure 38 shows the increased output capability of two and four loudspeakers, as compared with a single loudspeaker. Note that the reference response for a single loudspeaker has been normalized. That is, its response is assumed to be flat, whereas it will actually roll off below its resonance frequency.

EXAMPLE: Determine the frequency below which a pair of 18 inch subwoofers will mutually couple.

If the woofers are placed with their frames touching, their center-to-center distance will be 18 inches, or .46 meter. Thus,

$$f_0 = 339/(.46)\sqrt{2} = 339/.65 = 521 \text{ Hz}$$

References: Engebretson (1984)
Wolff and Malter (1929)

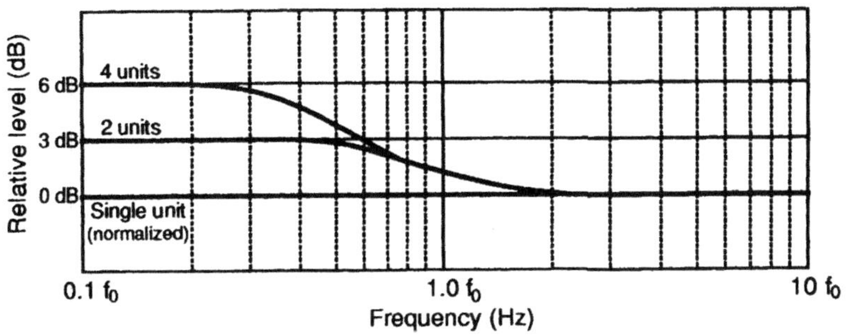

Figure 38. Normalized Mutual Coupling for Multiple Pistons

39. Acoustical Power Output Produced on One Side of a Piston in a Large Baffle as a Function of Amplitude, Radius, and Frequency

Figure 39 presents solutions to the following equation:

$$W_A = [x_{peak}\,(f^2a^2)]^2/2.32 \times 10^{17} \quad (39.1)$$

where x is the peak amplitude (mm), f is the driving frequency, and a is the piston radius (mm).

The calculations presented here are valid when the wavelength is large with respect to the piston diameter and small with respect to the baffle dimensions.

EXAMPLE: Determine the acoustical output power for a piston of radius 167 mm operating at a frequency of 52 Hz and a peak excursion of 2.3 mm.

First, determine the product fa = (52)(167) = 8684. Locate the x_{peak} value of 2.3 along the abscissa. Then read upward to the approximate value of 8684, which will be between the bold lines marked 8k and 9k. Then read the value of approximately 0.125 watts along the ordinate.

References: Beranek (1954)
Massa (1942)

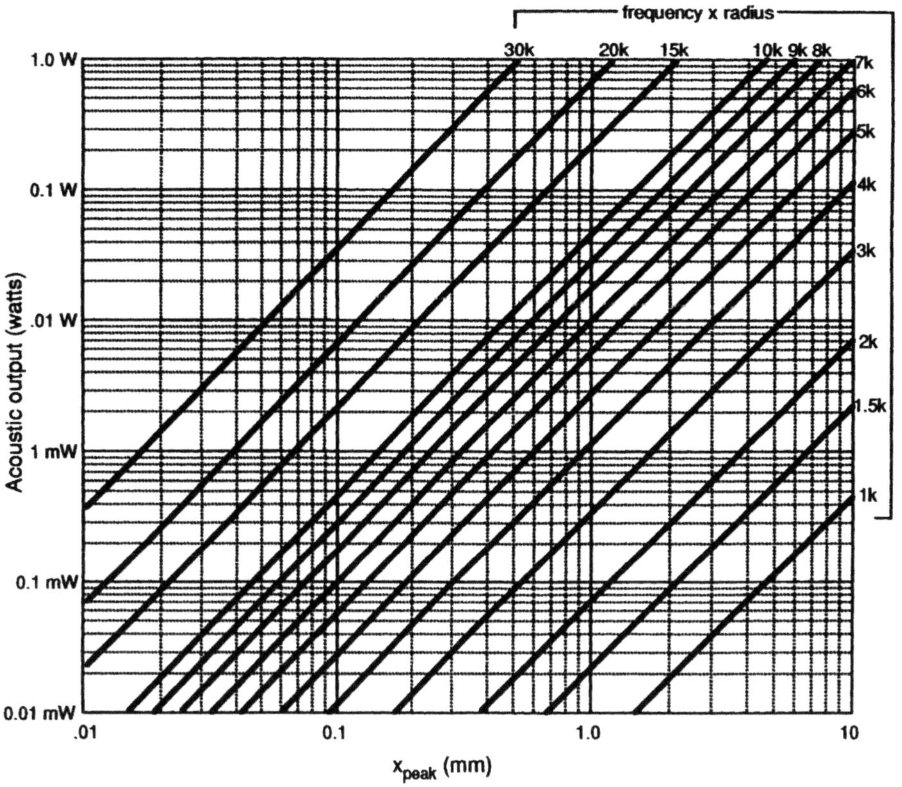

Figure 39. Acoustical Power Output Produced on One Side of a Piston in a Large Baffle as a Function of Amplitude, Radius, and Frequency

40. Sound Pressure Level Produced by a Piston in a Large Baffle at a Distance of 1 Meter as a Function of Amplitude, Radius, and Frequency

Figure 40 presents solutions to the following equation:

$$L_p = 20 \log [(x_{peak} (f^2 a^2))/1.18 \times 10^3] \quad \ldots\ldots\ldots\ldots\ldots\ldots \quad (40.1)$$

where x is the peak amplitude (mm), f is the driving frequency (Hz), and a is the piston radius (mm).

EXAMPLE: Determine L_p at a distance of 1 meter from a piston of radius 167 mm operating at a frequency of 52 Hz and peak excursion of 2.3 mm.

First, determine the product fa = (52)(167) = 8684. Locate the x_{peak} value of 2.3 along the abscissa. Then read upward to the approximate value of 8684, which will be between the bold lines marked 8k and 9k. Then read the value of 103 dB on the ordinate.

The calculations presented here are valid when the wavelength is large with respect to the piston diameter and small with respect to the baffle dimensions.

Reference: Gander (1981)

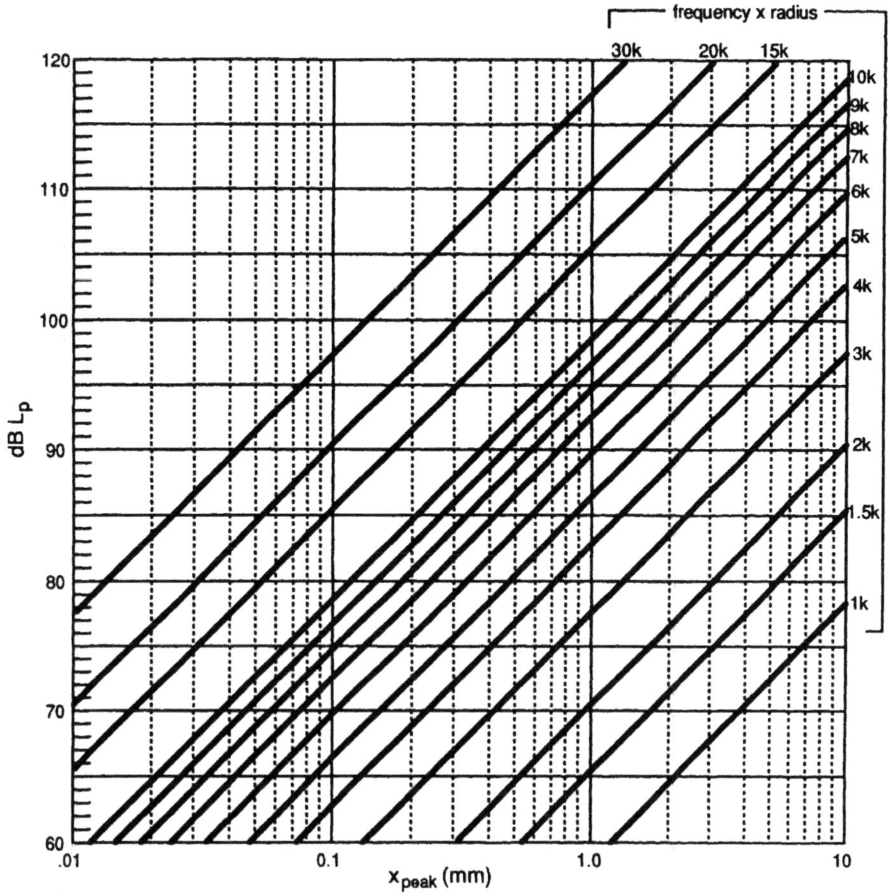

Figure 40. Sound Pressure Level Produced by a Piston in a Large Baffle at a Distance of 1 Meter as a Function of Amplitude, Radius, and Frequency

41. Sound Pressure Level Produced by a Piston in a Large Baffle as a Function of Radiated Power and Distance

Figure 41 presents solutions to the following equation:

$$L_p = 112 + 10 \log W_A - 20 \log r \quad \text{(41.1)}$$

where W_A is the radiated acoustical power and r is the measurement distance (m).

EXAMPLE: Determine the sound pressure level produced by a source radiating 20 acoustical watts at a distance of 10 meters from the baffle.

Locate 20 along the abscissa and read upward to the bold line that represents 10 meters. Then move across to the ordinate and read the value of 105 dB.

The calculations presented here are valid when the wavelength is large with respect to the piston diameter and small with respect to the baffle dimensions.

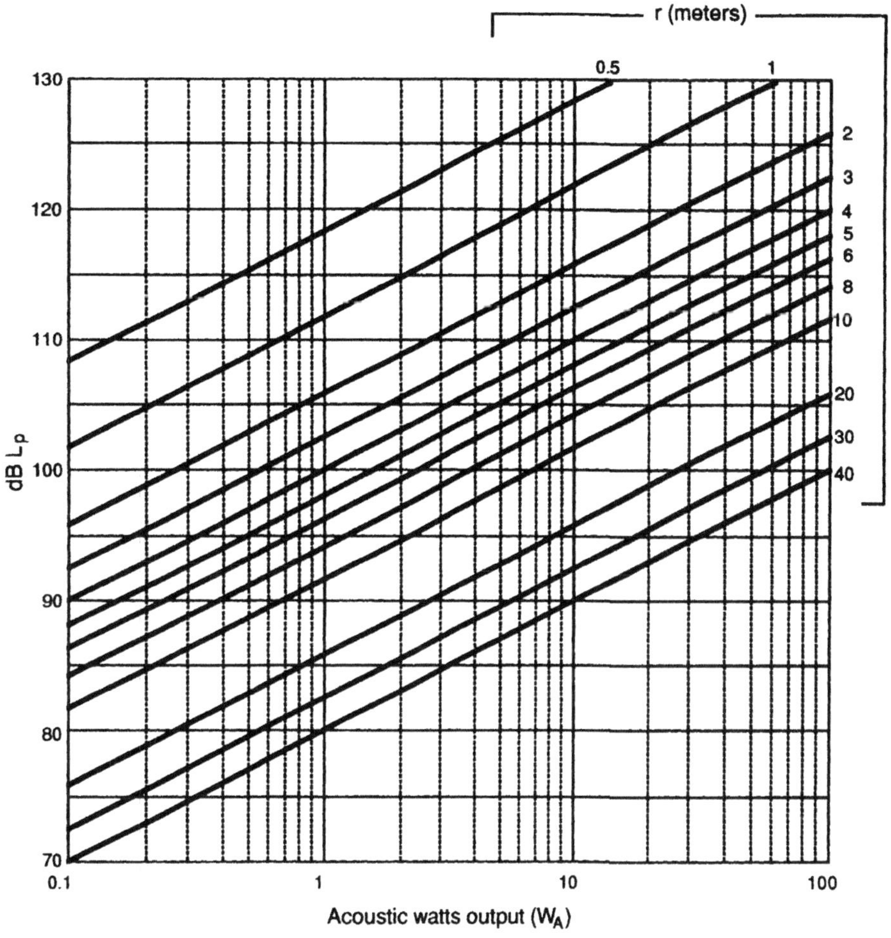

Figure 41. Sound Pressure Level Produced by a Piston in a Large Baffle as a Function of Radiated Power and Distance

42. PEAK AMPLITUDE FOR 1 ACOUSTICAL WATT RADIATED BY A PISTON INTO HALF-SPACE AS A FUNCTION OF RADIUS AND FREQUENCY

Figure 42 presents solutions to the following equation:

$$x_{peak} = (1.18 \times 10^{3})(10^{SPL/20})/f^2a^2 \quad (42.1)$$

where SPL has been fixed at 112 dB, which defines an acoustical power of 1 watt at a distance of 1 meter from a hemispherically radiating source. The peak displacement for any other power P may be determined by multiplying the solution for 1 watt by $\sqrt{P}$.

EXAMPLE: Find the peak amplitude of a 200-mm radius piston radiating 1 acoustic watt at 100 Hz.

Locate 100 Hz along the abscissa and 200 mm along the ordinate, then determine their intersection slightly above the 4-mm bold line. Estimate the value at about 1.2-mm peak amplitude.

EXAMPLE: Find the peak amplitude of a 500-mm radius piston radiating 2 watts at a frequency of 30 Hz.

Locate 30 Hz along the abscissa and 500 along the ordinate, then determine their intersection between the bold lines marked 2 and 3 mm. Estimate this value at 2.5 mm and multiply by $\sqrt{2}$, yielding 3.5-mm peak displacement.

References: Beranek (1954)
Gander (1981)
Massa (1942)

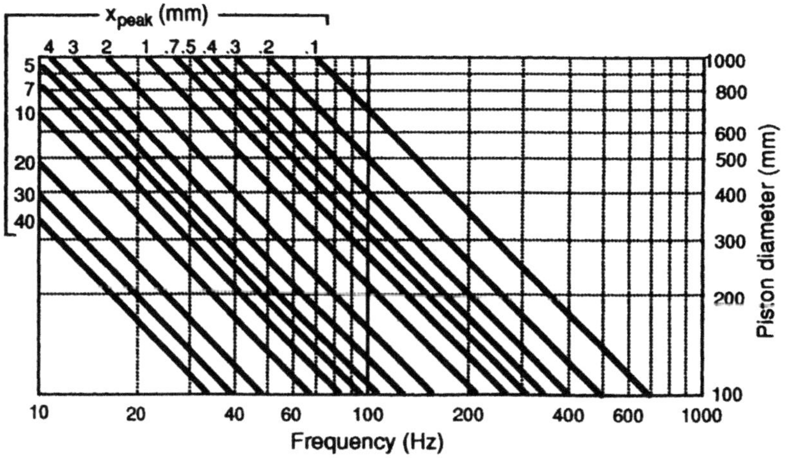

Figure 42. Peak Amplitude for 1 Acoustical Watt Radiated by a Piston into Half-Space as a Function of Radius and Frequency

43. Transducer Cone Deflection as a Function of Resonance Frequency

Transducers mounted in ceilings may sag over time. The distortion generated by those transducers will increase, because the voice coil is no longer centered in the magnetic gap. The amount of cone deflection is inversely proportional to the square of the resonance frequency of the moving system. It is generally believed that loudspeakers used in these applications should be chosen for as high a resonance frequency as possible.

The downward force on the cone is equal to Mg, where M is the mass of the cone in kilograms and g is the acceleration of gravity in meters per second squared. The cone is deflected downward until that force is balanced by the spring force of the suspension, ka, where k is the spring constant in newtons per meter, and a is the downward deflection in meters. Recalling that the equation for resonance is:

$$f_0 = \frac{1}{2\pi}\sqrt{\frac{k}{M}} \quad \text{(43.1)}$$

we can make the necessary substitutions and arrive at the following equation for deflection:

$$x = 247/f_0^2 \quad \text{(43.2)}$$

where x is the deflection in mm.

Figure 43 shows the range of cone deflection over a range of resonance frequencies from 10 to 100 Hz.

EXAMPLE: Find the deflection due to gravity of a loudspeaker cone with a resonance frequency of 50 Hz.

Locate 50 on the abscissa and read upward to the bold line. Then move over to the ordinate and read 0.1 millimeter.

References: Eargle (1989)
Olson (1957)

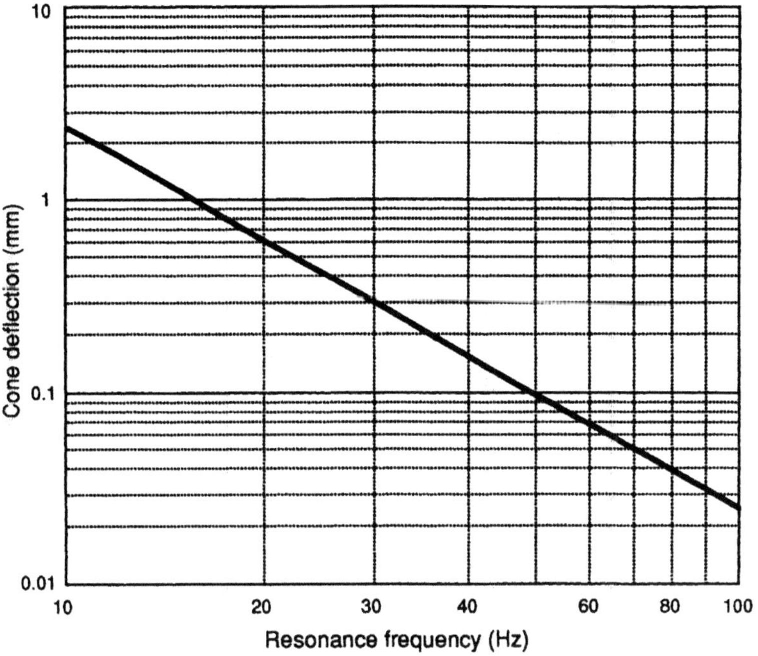

Figure 43. Transducer Cone Deflection as a Function of Resonance Frequency

44. SECOND HARMONIC DISTORTION IN HORNS

Figure 44 presents solutions to the following equation:

$$\%2\text{nd harmonic} = 1.76\sqrt{I_T}\,(f/f_c) \times 10^{-2} \quad \ldots\ldots\ldots\ldots\ldots (44.1)$$

where I_T is the acoustic intensity at the horn throat in watts per square meter, f_c is the nominal cutoff frequency of the horn, and f is the driving frequency. The graph, however, plots intensity in watts per square centimeter.

As a practical matter, it is sometimes difficult to arrive at the exact intensity at the throat of the horn. However, a good approximation may be made from a knowledge of the efficiency of the horn-driver combination.

The information in section 46 can be used to arrive at the efficiency of a horn-driver combination, as we will show in the following example:

EXAMPLE: Find the percent second harmonic distortion in a horn-driver combination with a cutoff frequency of 150 Hz and a driving frequency of 500 Hz. Assume that electrical power input to the driver is 10 watts and that the horn has a nominal sensitivity of 113 dB L_p at 1 meter and a directivity index of 11 dB.

First, refer to section 46. Find the DI along the abscissa and the sensitivity along the ordinate. Determine their intersection at the bold line corresponding to an efficiency of 20%. Multiplying the power input by the efficiency gives a net acoustic power of 2 watts.

Now determine the area of the throat of the horn. A good approximation here is to consider the horn as actually beginning at the phasing plug of the driver. In this case, assume that the driver has a 100-mm diaphragm and that the phasing plug has a 10-to-one area loading factor, which is normal for most high frequency drivers. The area of the diaphragm is:

$$\text{Area} = \pi(5)^2 = 78.4 \text{ square centimeters}$$

Taking one-tenth of this (the area loading factor of the phasing plug) gives 7.84 square centimeters.

The throat intensity can now be determined, as follows:

$$I_T = 2/(7.84) = .255 \text{ watts/cm}^2$$

The ratio f to f_c is 500/150 = 3.33. We can now go directly to Figure 44 to determine the distortion. Find the intensity of .225 along the abscissa and move upward to about halfway between the bold lines marked 2 and 4. Then move to the ordinate and read the value of about 2.5% distortion.

Reference: Beranek (1954)

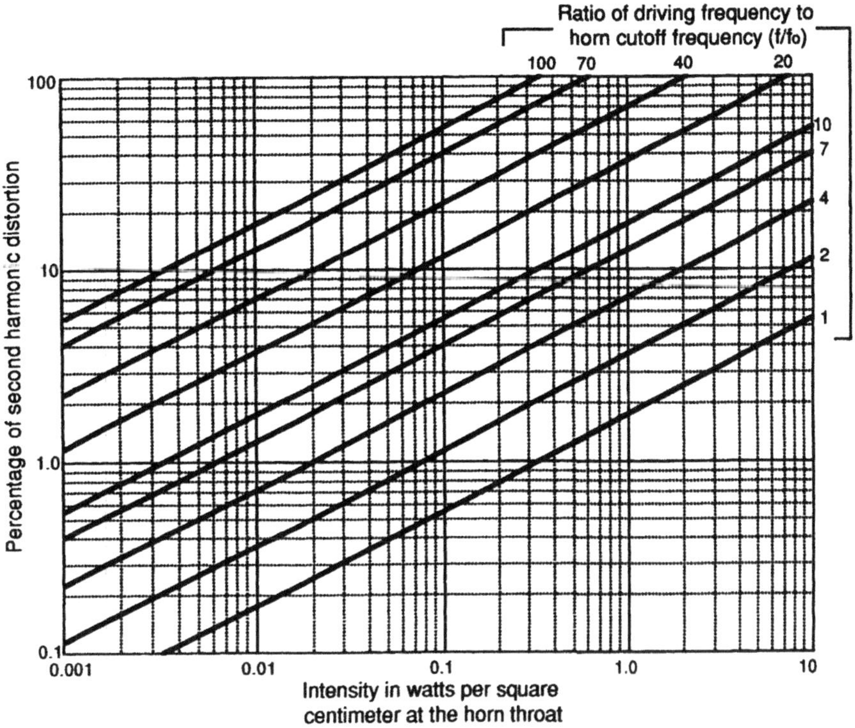

Figure 44. Second Harmonic Distortion in Horns

45. Frequency Modulation (FM) Distortion in Cone Transducers

Beers and Belar determined the total amount of FM distortion in a cone transducer to be:

$$\text{Percent distortion} = \frac{(1.86 \times 10^6)(f^2)\sqrt{W_1}}{(f_1^2)(d^2)} \quad \ldots\ldots\ldots\ldots\ldots (45.1)$$

where d is the diameter of the cone in millimeters, f_1 is the modulating frequency, W_1 is the acoustical power output at f_1, and f_2 is the modulated frequency. The values of distortion given in this example will be apparent only along the axis normal to the cone.

As a rule, the bulk of the power output from the transducer is at a lower modulating frequency, f_1. F_2 is normally a much higher frequency than f_1, and the amount of power radiated at f_2 is generally negligible. Thus, f_2 will be frequency modulated by f_1, and the amount of distortion is relatively independent of the amplitude of f_2.

FM distortion comes into discussion any time a low-frequency transducer in a system is forced to work at fairly high frequencies. For example, a 381-mm (15-in) diameter low-frequency transducer is normally limited to about 400 or 500 Hz in its operation; but in many coaxial loudspeakers, its response is carried up to about 1200 Hz. Figure 45 illustrates this case for a modulating frequency of 100 Hz.

EXAMPLE: A 381-mm diameter transducer is generating a 100-Hz signal at an acoustical power output of 4 watts. What will be the percentage FM distortion of a 1250-Hz signal superimposed on it?

Locate f_2 at 1250 Hz along the abscissa and move upward to the bold line marked 4 watts. Then move to the ordinate and read the value of about 3% FM distortion.

For reference, the output power of 4 watts at f_1 corresponds to a level at 1 meter from the loudspeaker of 115 dB L_p.

Reference: Beers and Belar (1943)

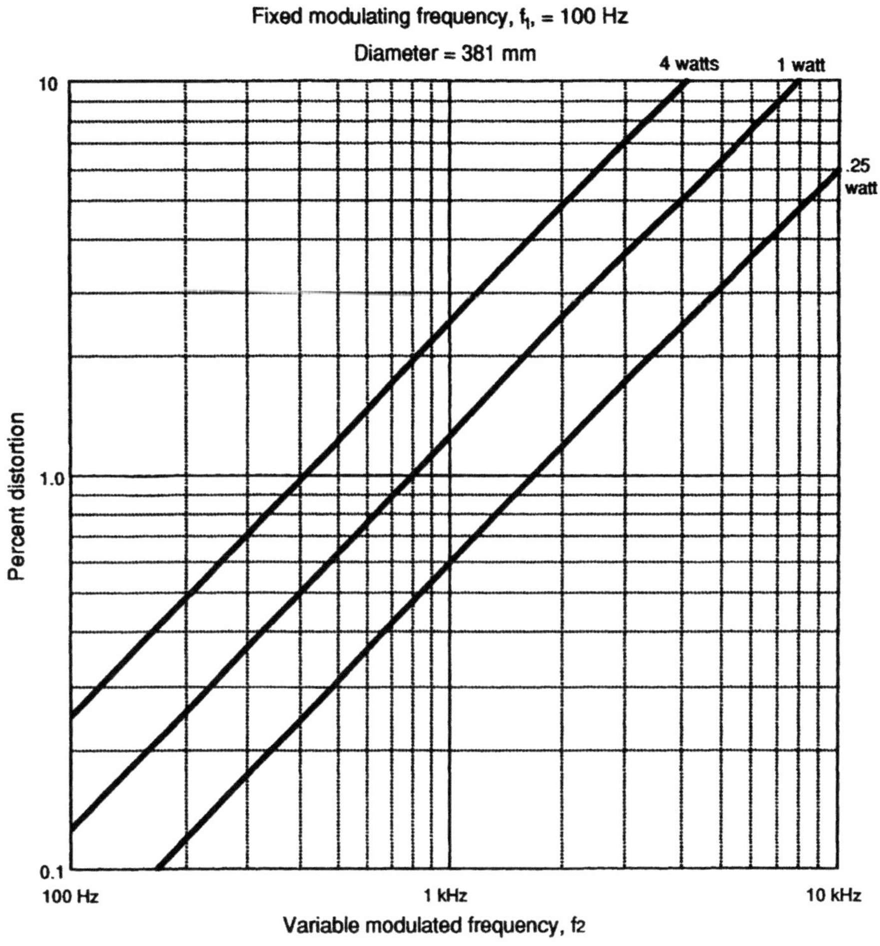

Figure 45. Frequency Modulation (FM) Distortion in Cone Transducers

46. Nominal Loudspeaker Efficiency as a Function of On-Axis Sensitivity and Directivity Index

Figure 46 presents solutions to the following equation:

Sensitivity (1 W@ 1 m) = 109 + DI + 10 log (Efficiency) . . . . (46.1)

The equation itself is based on the fact that 1 acoustic watt radiating uniformly in free space will, at a distance of 1 meter, produce L_p of 109 dB. This can be seen in equation 17.2 by setting Q, W_A, and r equal to unity and letting R approach infinity.

We use the term "nominal," since this implies that the impedance of the loudspeaker is constant. We know that this is not actually the case, but we assume it to be for the sake of simplification and because it creates a relative lack of confusion in loudspeaker-amplifier specifications in the field.

EXAMPLE: At a specified frequency, a loudspeaker has an on-axis sensitivity value of 105 dB, a 1-watt input as measured at a distance of 1 meter, and a calculated value of on-axis directivity index of 10 dB. Find the nominal efficiency of the loudspeaker.

Locate the directivity index and sensitivity values on their respective axes and determine their intersection point. Then read the value of efficiency from the closest bold line. In this case, the efficiency is 4%.

It is of utmost importance to ensure that published sensitivity values have been referred to the nominal impedance of the loudspeaker. For example, if the loudspeaker has a nominal impedance of 8 ohms, this implies that the modulus of impedance reaches a value of perhaps 7 ohms slightly above the primary resonance, but it is 8 ohms or greater elsewhere. In this case, an applied signal of 2.83 volts rms is used to derive the nominal 1-watt input referred to in this section. See the following section for more details.

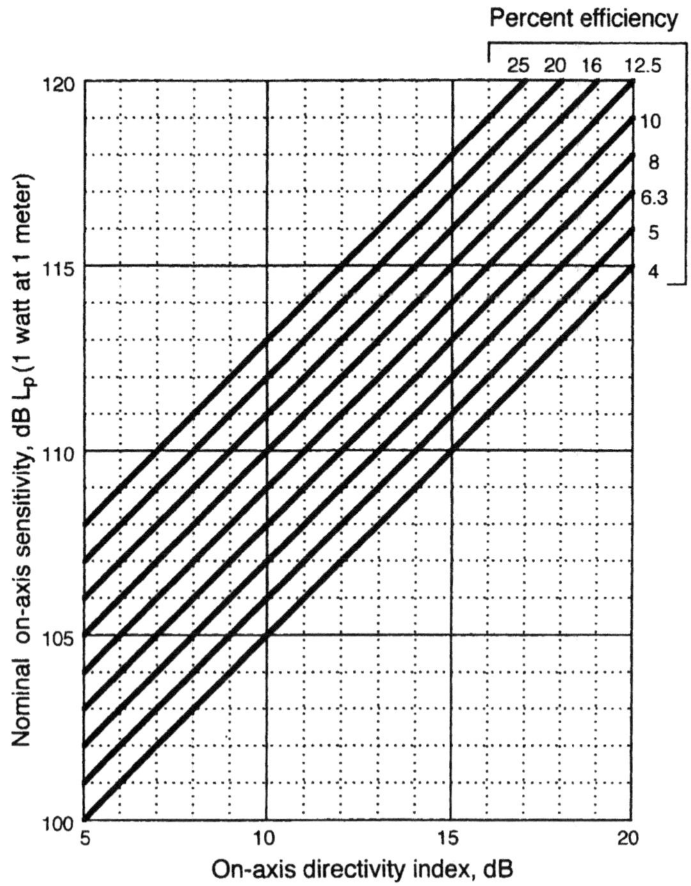

Figure 46. Nominal Loudspeaker Efficiency as a Function of On-axis Sensitivity and Directivity Index

47. Sensitivity Ratings for Loudspeaker Systems

Figure 47 gives various sets of data related to the often confusing matter of loudspeaker sensitivity. Generally, power sensitivity is implied, unless voltage sensitivity is clearly stated.

If this is the case, then the applied voltage is dependent on the loudspeaker's nominal impedance, as given in the chart in Figure 47A. The acoustical measurements themselves may be stated variously. Today, the standard reference is 1-watt at 1-meter, but a few manufacturers will give additional data of 1 watt measured at 4 feet, which should read 1.7 dB less than the 1-meter rating. The EIA method, which is rarely encountered today, gives a value that is about 49 dB lower than the 1-watt, 1-meter method.

Voltage sensitivity is another matter. It is largely an outgrowth of the consumer electronics industry, where inflated ratings are relatively commonplace. The normal tendency here is to give a rating based on an input signal of 2.83 volts rms, even though the loudspeaker may have an impedance closer to 4 ohms than 8. From Figure 47B, we can see that, in this case, the rating will be 3 dB higher than it would be if a normal power rating was presented. Let the user beware.

A Power sensitivity:

One watt into nominal impedance Z_{nom}:

Based on the equation:

$$E_{in}^2 = Z_{nom}$$

Z_{nom}	E_{nom}
16	4 V
12	3.46
8	2.83
6	2.45
4	2

Acoustical measurements:

L_p (1 W @ 1 m) = 0 dB reference (preferred)
L_p (1 W @ 4 ft) = -1.7 dB (older English units)
L_p (1 mW @ 30 ft) = -49.2 dB (EIA method)

B Voltage sensitivity:

2.83 V@ 1 m:

Z_{nom}	Level[1]
8 ohms	0 dB reference
4 ohms	+3 dB
16 ohms	-3 dB
6 ohms	+1.25 dB
12 ohms	-1.76 dB

[1] Level given by the following equation:

$$\text{Level (dB)} = 0 - \log(Z_{nom}/8)$$

Figure 47. Sensitivity Ratings for Loudspeaker Systems

48. Plane Wave Tube (PWT) Sensitivity Ratings for Compression Drivers

Plane wave tube measurements require careful reading on the part of the user, inasmuch as a number of standards are to be found in the field. However, professional manufacturers often present detailed specifications, and all pertinent information is usually given somewhere in the specification.

A plane wave tube is universally used to derive the absolute output of a compression driver as a function of frequency. It consists of a rigid tube, ideally the same diameter as the driver's exit. It is gradually damped along its length, so that acoustical power is effectively absorbed before reaching the end of the tube. It is often referred to as a "terminated tube." A microphone embedded in the wall of the tube close to the driver is used to sample the pressure in the tube.

At one time, all manufacturers presented PWT measurements based on 1 milliwatt input to the tube. Under this condition, there would be a difference in measured L_p in the tube based on the cross-sectional area of the tube; this information is given in Figure 48A. Some manufacturers rescaled their 2-inch tube data so that it matched the equivalent 1-inch data.

There is a recent trend in the industry to state PWT sensitivity data based on a reference value of intensity of 1 mW/cm^2. These data are give in Figure 48B.

All of these matters grow more complex with the proliferation of compression drivers with different exit diameters. The great virtue of this new rating method is that, for a given driver efficiency, the L_p in the tube will be constant, regardless of its diameter. This can easily be seen in Figure 48B.

EXAMPLE: One manufacturer rates a 51-mm exit driver based on a measurement made on an 18-mm (.75 in) PWT and states that the 1-mW sensitivity is 118.5 dB L_p. Another manufacturer of a 51-mm exit driver uses a 51-mm PWT for making its 1-mW measurements and states that the sensitivity is 116 dB L_p, referred to a 25.4-mm (1-in) tube. How do we compare the drivers?

In Figure 48A, note that 116 dB on the 25.4-mm tube represents the same efficiency (20%) as the value of 118.5 dB measured on the 18-mm tube. Thus, the two drivers are equivalent in sensitivity at the stated frequency.

A

For 1 mW electrical power input to driver

18 mm (.75 in)	25.4 mm (1 in)	51 mm (2 in)	76 mm (3 in)	Percent efficiency of driver
125.5 dB	123 dB	117 dB	113.5 dB	100
124.5	122	116	112.5	80
123.5	121	115	111.5	63
122.5	120	114	110.5	50
121.5	119	113	109.5	40
120.5	118	112	108.5	31
119.5	117	111	107.5	25
118.5	116	110	106.5	20
117.5	115	109	105.5	16
116.5	114	108	104.5	12.5
115.5	113	107	103.5	10
114.5	112	106	102.5	8
113.5	111	105	101.5	6.3

Diameter of plane wave tube (PWT)

B

Electrical power input required for 1 mW per square cm

2.8 mW	5 mW	20 mW	45 mW	
130 dB	130 dB	130 dB	130 dB	100
129	129	129	129	80
128	128	128	128	63
127	127	127	127	50
126	126	126	126	40
125	125	125	125	31
124	124	124	124	25
123	123	123	123	20
122	122	122	122	16
121	121	121	121	12.5
120	120	120	120	10
119	119	119	119	8
118	118	118	118	6.3
18 mm (.75 in)	25.4 mm (1 in)	51 mm (2 in)	76 mm (3 in)	Percent efficiency of driver

Diameter of plane wave tube (PWT)

Figure 48. Plane Wave Tube (PWT) Sensitivity Ratings for Compression Drivers

49. Radiation Resistance for Various Horn Flare Development Curves

Three possible horn flares are shown in Figure 49A, and their relative responses are shown in Figure 49B. The conical horn (A) provides the least resistive loading in the region of horn cutoff at 100 Hz, whereas the hyperbolic flare provides the greatest. Because of its rapid flare rate, the conical horn provides the least distortion (see section 44). The hyperbolic horn (C) provides the highest distortion because of its relatively slow increase in cross-sectional area along the horn's axis.

Traditionally, the exponential horn (B) has been used in horn design because of its good loading down to the cutoff frequency and because it is a reasonable compromise between the low distortion of the conical horn and the high distortion of the hyperbolic horn.

Reference: Beranek (1954)

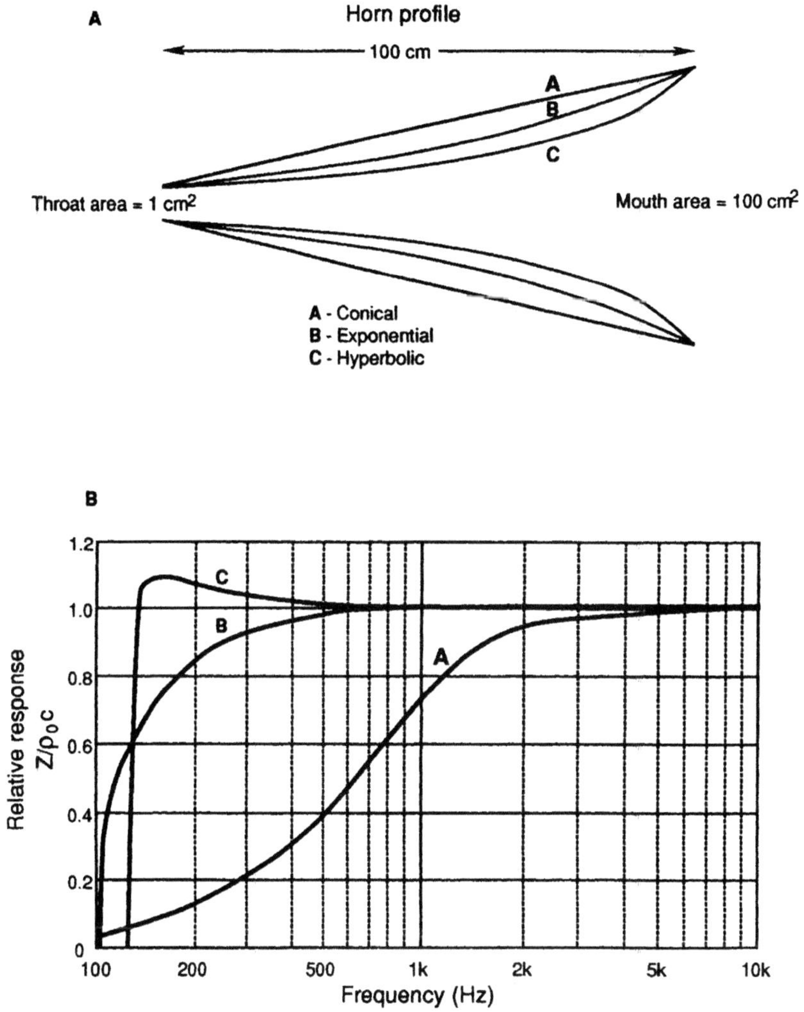

Figure 49. Radiation Resistance for Various Horn Flare Development Curves

50. High-Frequency Driver Electrical Derating for Flat Power Response Equalization

The typical plane wave tube response of a compression driver is given in Figure 50. The response at low frequencies takes place below the primary resonance of the device, and it determines the lower usable frequency of the device. At higher frequency, normally in the 3–4.5 kHz range, there is another inflection point in the curve, known as the "mass breakpoint," above which response falls at 6 dB per octave.

The maximum power input the driver can accommodate is called its thermal power rating; this is indicated on the ordinate of Figure 50A as $W_{thermal}$. Assuming that we want to accommodate a flat input signal, it is clear that the driver's output will be rolled off above the mass breakpoint. If we wish to maintain flat response beyond the mass breakpoint out to some f_1, it will be necessary to boost the signal as shown. When this is done, the loudspeaker's output will be flat out to f_1 and will fall off above this frequency at 6 dB per octave.

If we want to accommodate a flat input signal, we will have to limit the total power input to some lower amount. Figure 50B gives the amount of derating required for various amounts of input signal boost that we may want to maintain.

In most professional applications involving uniform coverage horns, such power response equalization is necessary. The maximum amount of boost is normally about 10 dB; this is sufficient to make the system's power response flat to 10 kHz.

Reference: Eargle (1989)

A

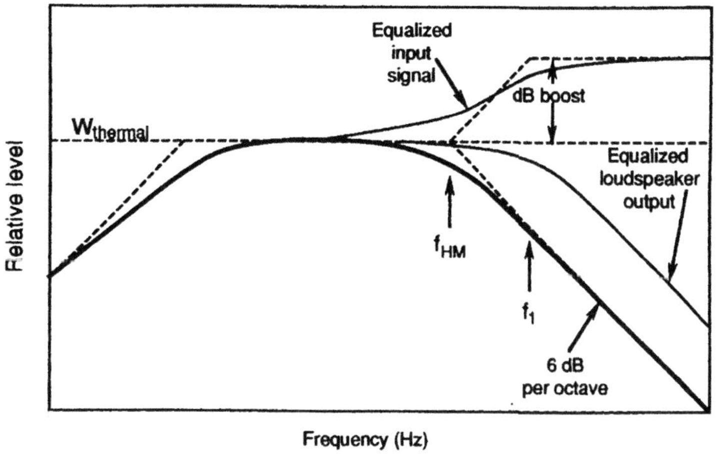

B

dB boost	Derating factor
3	0.5
6	.25
10	0.1

Figure 50. High-Frequency Driver Electrical Derating for Flat Power Response Equalization

51. Duty Cycle–Related Power Ratings

In addition to a thermal rating, most transducers carry a displacement power rating that varies with frequency. The thermal rating is concerned only with heating effects; it is constant throughout the frequency range. The displacement power rating is excursion-related; it tends to fall at lower frequencies, where the transducer's cone or diaphragm motion increases.

At higher frequencies, the device can usually handle short bursts of input power higher than the thermal rating, but only if the displacement rating is not exceeded in the process. Furthermore, the duty cycle for the increased input must be such that the transducer has time to dissipate the increased thermal load. Power input increases of up to three to one are possible if the duty cycle is short enough so that the driver's voice coil has time to cool down to a safe temperature.

It is difficult to establish the boundary between thermal and displacement power ratings for enclosed cone transducers, due to the vast differences encountered in enclosure tuning. But in the case of compression drivers, it is often easy to determine the boundary.

The first thing to note here is the manufacturer's recommended lower crossover frequency. This is normally close to the boundary we are looking for, and it can be plotted on the driver's plane wave tube curve as a 6-dB-per-octave curve, as shown in Figure 51. The transition between thermal and displacement ratings is indicated as f_1 in the figure.

Thus, in the octave above the lowest recommended crossover frequency, it may be possible to operate the driver at twice the thermal rating, but only if a two-to-one duty cycle can be ensured. Following the slope upward, we can identify a region over which a three-to-one power increase and corresponding duty cycle can be identified.

We strongly recommend that manufacturers be consulted before an operation such as the one described here is implemented. In most cases, manufacturers will have helpful suggestions for the user.

Reference: Eargle (1989)

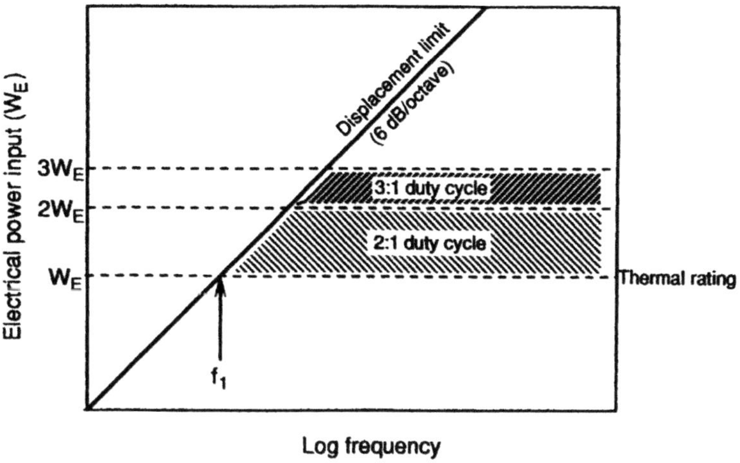

Figure 51. Duty Cycle–Related Power Ratings

52. Resistance Change with Temperature for Copper

The resistance of a conductor is given by the following equation:

$R = R_0(1 + \alpha t)$ . . . . (52.1)

where R_0 is the resistance at 0°C, α is the temperature coefficient, and R_0 is the resistance at temperature t°C. For copper, the value of α is .00393°C^{-1}.

In Figure 52, we have assumed a 10-meter length of AWG no.10 copper wire (with a cross-sectional area of 5.2 square mm).

EXAMPLE: A given transducer has a copper voice coil with a dc resistance of 7 ohms at 25°C. Find the resistance at the elevated temperature of 125°C.

Find 25°C on the abscissa. Move up to the bold line and read a value of .036. Then, locate 125°C on the abscissa and read the value of about .049. The value of resistance will then be:

Resistance = 7(.049/.036) = 7.53 ohms

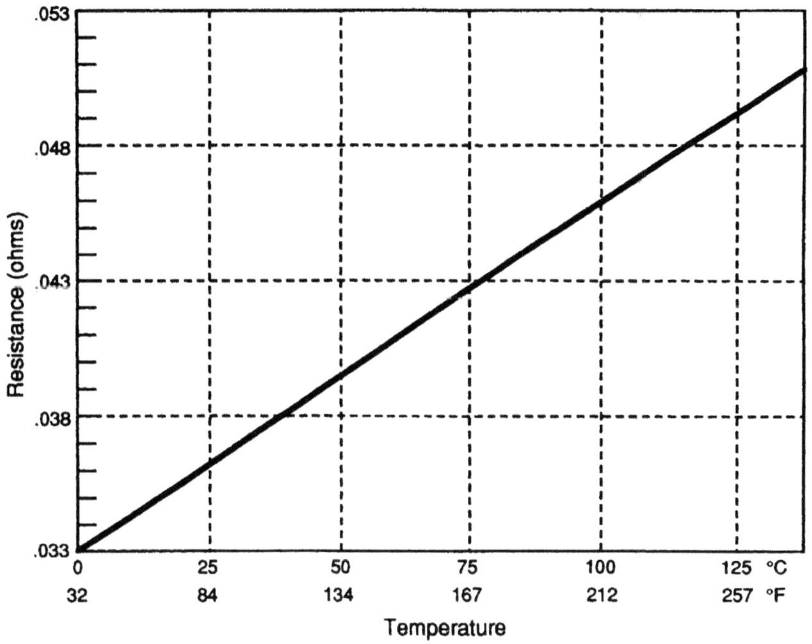

Figure 52. Resistance Change with Temperature for Copper

53. Weighting Curves for Loudspeaker Power Measurements

The problem here is to establish a standard test signal, easily implemented in the field, that corresponds to real-world usage so that manufacturers can reasonably—and competitively—specify their products in terms of nominal power handling.

Most music and speech spectra exhibit rolled-off response at very high and very low frequencies; loudspeaker systems may accordingly be designed to handle less sustained input power at those frequency extremes. A fair estimate of a given loudspeaker's ability to handle program power may be determined by studying its various failure modes as an input noise signal, suitably shaped, is applied at increasing power increments.

Curve 1 of Figure 53A has been suggested by the Electronic Industries Association of America (EIA) as an appropriate weighting curve for this purpose. The input signal to the weighting filter is white noise. No tolerance is stated. Curve 2 has been proposed by the EIA as an update for curve 1 that will take into account the demands of today's program material.

The data shown at Figure 53B have been suggested by the International Electrotechnical Commission (IEC) for the same purpose. Here the upper and lower limits of the white noise spectrum are given. Today, the IEC weighting curve is in general use, since it corresponds more accurately to the broad range of spectra found in contemporary music; however, the proposed EIA curve will fulfill basically the same function.

For determining the power rating of individual transducers that may be specified in systems, the Audio Engineering Society (AES) has suggested the guidelines shown at Figure 53C. For this specification, the manufacturer defines a frequency decade over which the transducer may be used; and pink noise is introduced over that decade, with high- and low-frequency rolloff of 12 dB per octave. A given transducer may carry more than one power rating, depending on the particular frequency decade used for the measurement.

For all three standards presented here, the input noise signal is preconditioned so that the peak-to-rms voltage value is four to one (6 dB).

References: EIA Document RS-420-A (1980)
IEC Document 268–5 (1977)
AES Document AES2 (1984)

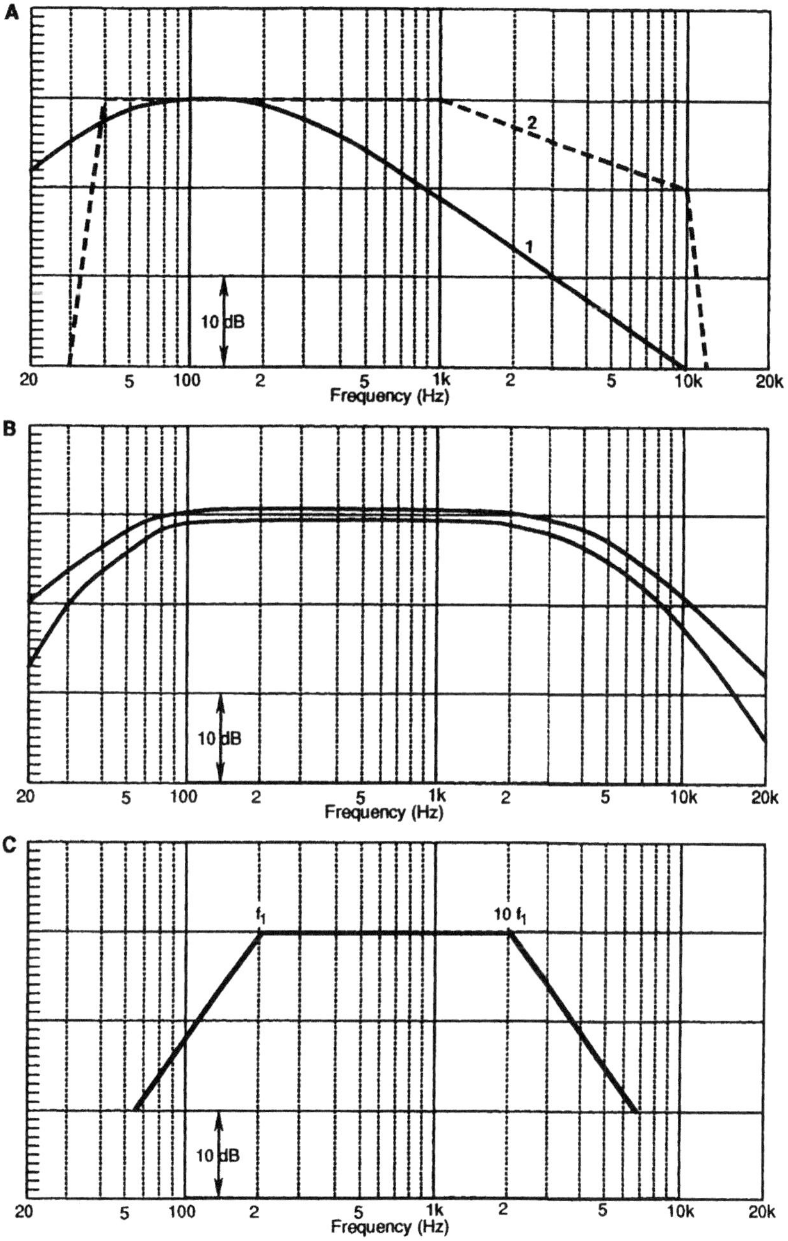

Figure 53. Weighting Curves for Loudspeaker Power Measurements

54. House Equalization Standard Curves for Sound Reinforcement and Program Monitoring

There are no standard "house curves" for monitor systems in recording studios; however, a number of recording companies and specialists in control room design have established their own equalization contours.

The curve shown at Figure 54A is typical of what may be expected today in a modern control room. The response can normally be maintained to a tolerance of ±2 dB from about 100 Hz to the upper limit. A more important concern is that the two stereo channels be very closely matched; if the control room is structurally symmetrical and the loudspeaker components flush mounted, it should be possible to match the two channels within 1 dB of each other, as measured on one-third octave centers. Under conditions of architectural symmetry, both left and right equalizers should have virtually the same settings.

The curve shown at Figure 54B has been widely applied to speech reinforcement systems. Paul Boner was probably the first acoustician to recommend its use. Response below 100 Hz and above 10 kHz is normally allowed to roll off, as shown.

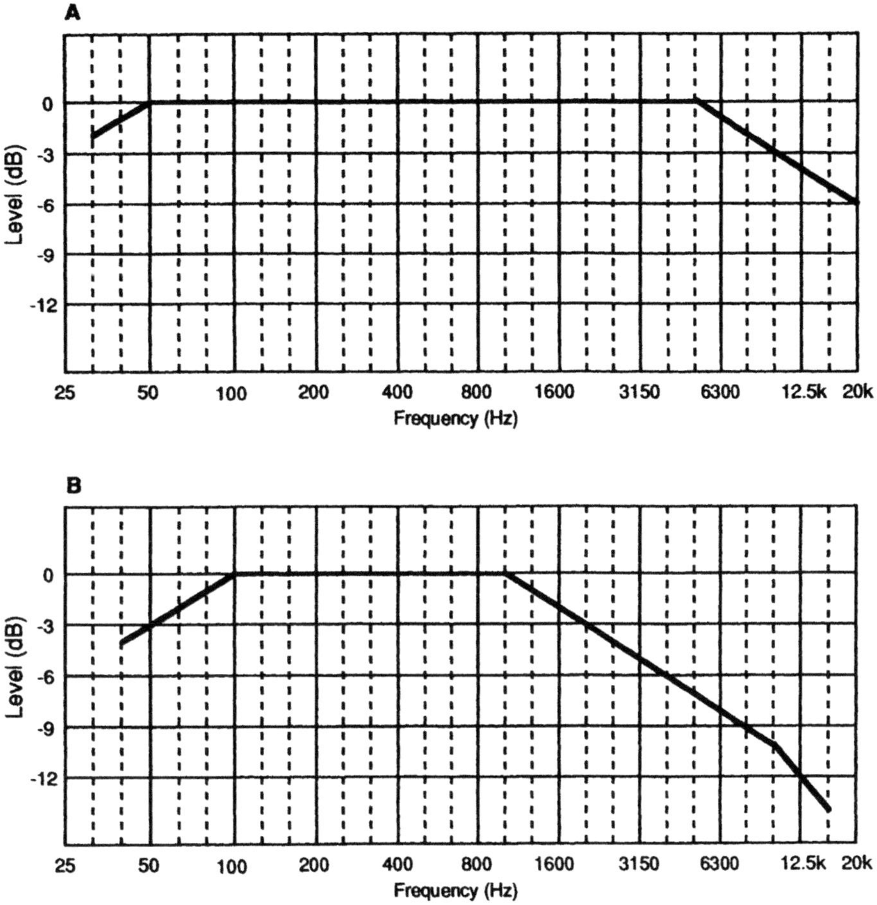

Figure 54. House Equalization Standard Curves for Sound Reinforcement and Program Monitoring

55. Transducer Sensitivity as a Function of Atmospheric Pressure and Temperature

As indicated in Figure 55, there is relatively small variation in transducer sensitivity observed over the normal range of temperature and atmospheric pressure. However, in making sensitive acoustical measurements, many engineers and acousticians consider both factors and make allowances for them.

The loss is given by the following equation:

$$\text{Loss} = 10 \log\left[\sqrt{\frac{°\text{F} + 460}{527}}\left(\frac{30}{\text{B}}\right)\right] \quad \ldots\ldots\ldots\ldots\ldots\ldots\ldots (55.1)$$

where B is the air pressure in inches of mercury and °F the temperature in degrees Fahrenheit.

At sea level, B is approximately 30, and we can see that the loss between temperatures of 20°C and 49°C is about 0.4 dB. Values of B of 25 inches and 20.5 inches correspond to altitudes of 1500 and 3000 meters, respectively.

Reference: Beranek (1954)

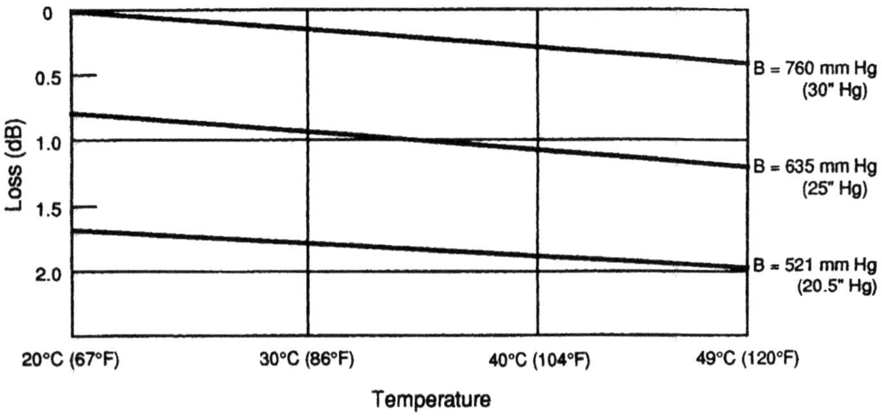

Figure 55. Transducer Sensiivity as a Function of Atmospheric Pressure and Temperature

56. Relation between 2π and 4π Loading and Baffle Size

A transducer mounted in a baffle will radiate effectively into half-space (2π loading) at frequencies above which the baffle perimeter is equal to two or more wavelengths. At longer wavelengths, the loudspeaker begins to radiate into full-space (4π loading).

The basic relationship is shown in section 36, where it can be seen that, above ka = 2, the DI increases 6 dB per doubling of frequency. In the example given in this section, we are effectively relating the baffle perimeter directly to ka.

As a general rule, the dimensions *a* and *b* should be of a relatively small ratio—perhaps no greater than 1.6—and the transducer should be located toward the middle of the baffle.

EXAMPLE: A loudspeaker is placed in a rectangular baffle that measures 1 by 1.5 meters. Above what frequency will radiation from the loudspeaker be essentially into 2π space?

Determine that the perimeter of the baffle is 5 meters. Locate 5 meters along the abscissa of the graph and move upward to the bold line. Then, read approximately 150 Hz along the ordinate.

Reference: Beranek (1954)

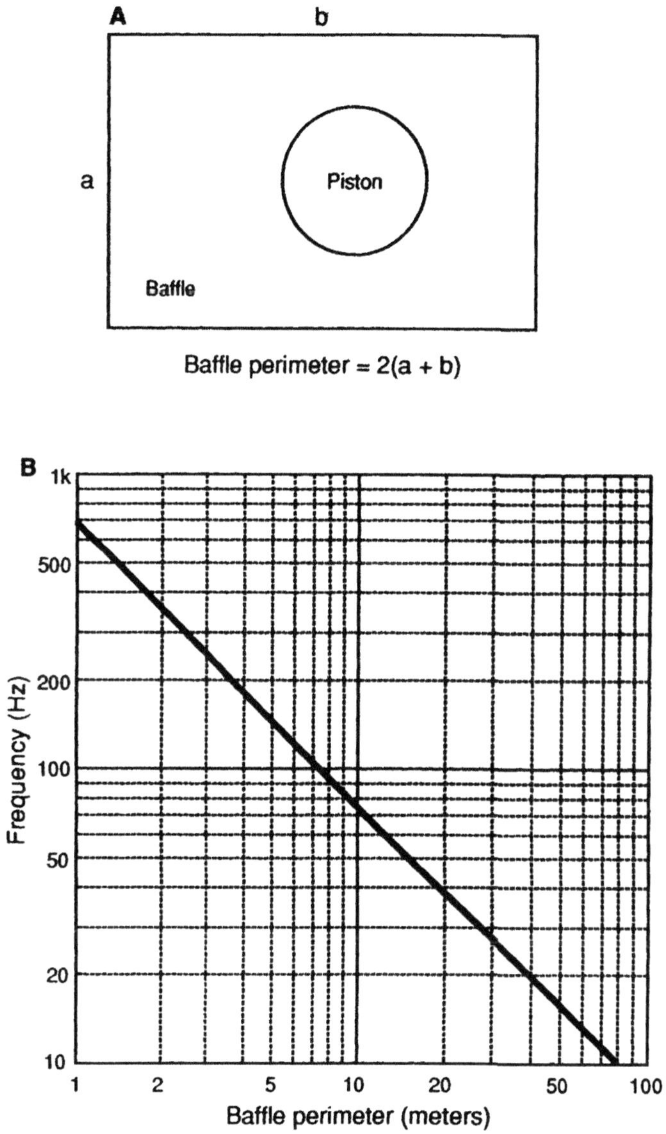

Figure 56. Relation between 2π and 4π Loading and Baffle Size

57. Horn Mouth Size versus –6 dB Beamwidth Control

The data presented at Figure 57A show the approximate beamwidth behavior of a typical 90° × 40° high-frequency horn. We can identify clearly the frequencies at which the nominal –6 dB horizontal and vertical beamwidths have doubled. The frequencies at which this occurs can be easily estimated from the horn mouth dimension in the plane we are concerned with.

It has been observed that horns with rectangular apertures exhibit a consistent loss of pattern control as given by the following table:

Nominal coverage angle	*Nominal angle doubles at*
90°	λ = 3 times mouth height
60°	λ = 2 times mouth height
40°	λ = 4/3 times mouth height
20°	λ = 2/3 times mouth height

These four relationships have been plotted at Figure 57B. The range of useful pattern control begins about 1 octave above the frequency at which the pattern has doubled.

EXAMPLE: Find the required mouth size for a horn with nominal 20° × 20° pattern control that extends downward to 1 kHz.

For useful pattern control at 1 kHz, the horn's pattern control will have doubled at about 500 Hz. Locate 500 Hz on the abscissa and read upward to the bold line marked 20°. Then read 1000 mm on the ordinate. The mouth will be approximately a square 1 meter on a side.

This section underscores the extreme space requirements for horns that have narrow pattern control at low frequencies.

Reference: JBL (1984)

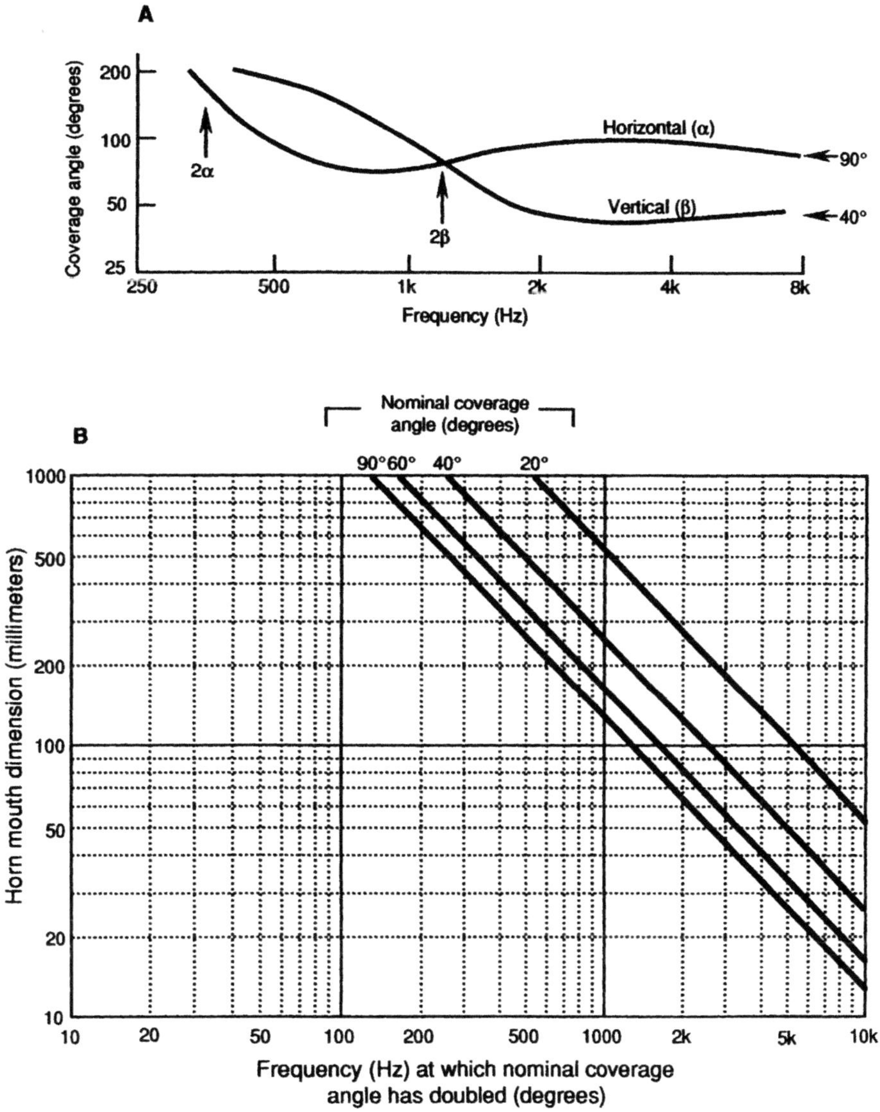

Figure 57. Horn Mouth Size versus –6 dB Beamwidth Control

58. BEAMWIDTH CONTROL OF MULTICELLULAR HORNS

The data presented in Figure 58 show the theoretical –6 dB beamwidth control of multicellular horns as a function of the number of cells in one plane. Note that for horns consisting of more cells, the useful frequency range extends lower. Just below that useful limit, the pattern control narrows greatly. This takes place at the frequency whose wavelength is approximately equal to the distance spanned by the group of cells. Below the frequency of narrowing, the pattern control becomes increasingly wide.

These data give a fair approximation for the behavior of multicellular horns at frequencies below, say, 3 or 4 kHz. At higher frequencies, the directional response of these devices may become erratic, with pronounced lobing along each septum that divides the adjacent cells. Multicellular horns are rarely specified today because of this problem.

Reference: Beranek (1954)

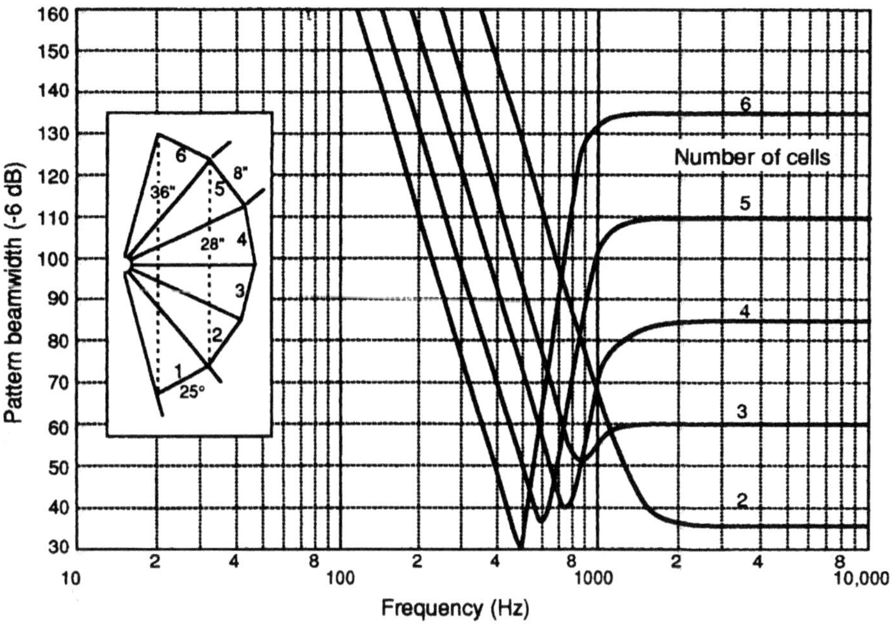

Figure 58. Beamwidth Control of Multicellular Horns

59. Beamwidth Narrowing with Vertical Stacked Horn Arrays

The data shown in Figure 59 are intended to offer general guidelines in vertical stacking of horns. The example used here is the JBL model 2370 horn, which has a well-controlled horizontal beamwidth of 90°. The vertical beamwidth response, due to the fairly small vertical mouth dimension, is broad at lower frequencies and narrows progressively with rising frequency.

When the horns are stacked vertically, the horizontal beamwidth remains unchanged, but the vertical beamwidth becomes narrower at lower frequencies due to the larger effective mouth size.

Horn stacking is fairly predictable up to three units; above that number, there may be problems in interference and lobing due to the large vertical dimension of the array. In general, we recommend horn stacking only when the overall vertical dimension can be kept fairly small.

With the horn discussed here, the substantial vertical narrowing in the 2-kHz octave range can be quite useful in speech reinforcement applications through the increase in direct-to-reverberant response in that frequency range.

References: Eargle (1989)
JBL (1984)

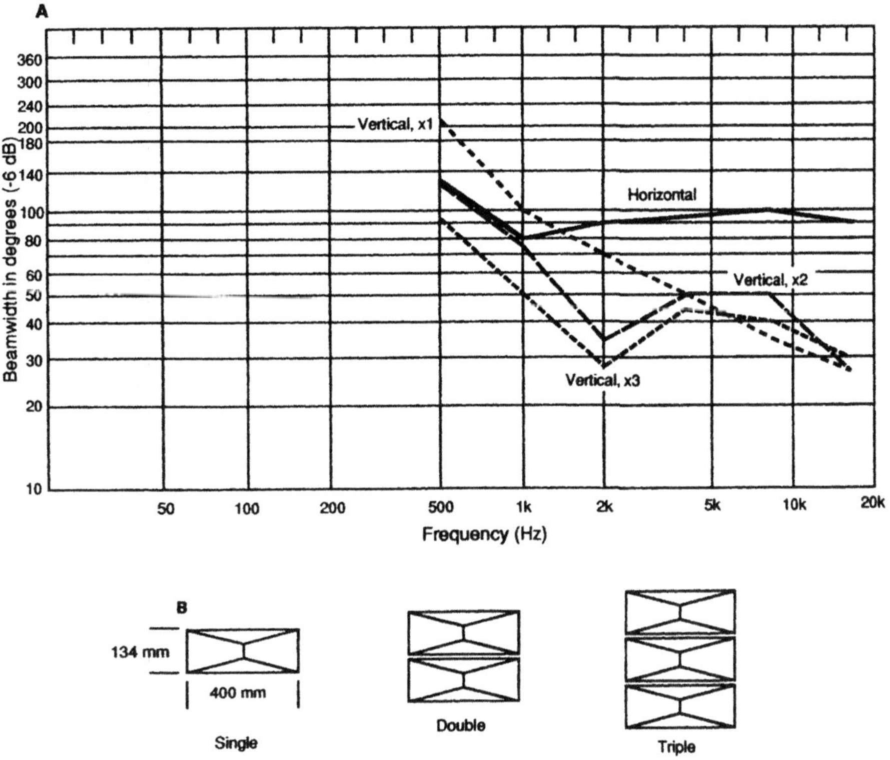

Figure 59. Beamwidth Narrowing with Vertical Stacked Horn Arrays

60. Directivity versus Horizontal and Vertical Beamwidth

The directivity properties of many horns can be quickly determined from the nomograph shown in Figure 60. The nomograph is based on the following equation:

$$DI = 10 \log \left[\frac{180°}{\arcsin \left(\sin \frac{\alpha}{2} \bullet \sin \frac{\beta}{2} \right)} \right] \quad \ldots\ldots\ldots\ldots\ldots\ldots (60.1)$$

The assumption is made that the horn's response falls off rather rapidly outside the nominal –6 dB beamwidth; this is a reasonable assumption at mid and high frequencies. Note carefully, however, that many horns are specified to operate at frequencies below this point, and their beamwidth may not fall off rapidly at angles wider than the specified –6 dB points.

EXAMPLE: Find the DI of a horn that has nominal 90° × 40° coverage.

Place a straightedge so that 90° is intersected along the α column and 40° along the β column. Then read the value of 11 dB along the middle column. For comparison, the JBL model 2360 horn, with nominal 90° × 40° coverage, has a measured DI of 10.8 dB averaged from 500 Hz to 16 kHz.

Reference: Molloy (1948)

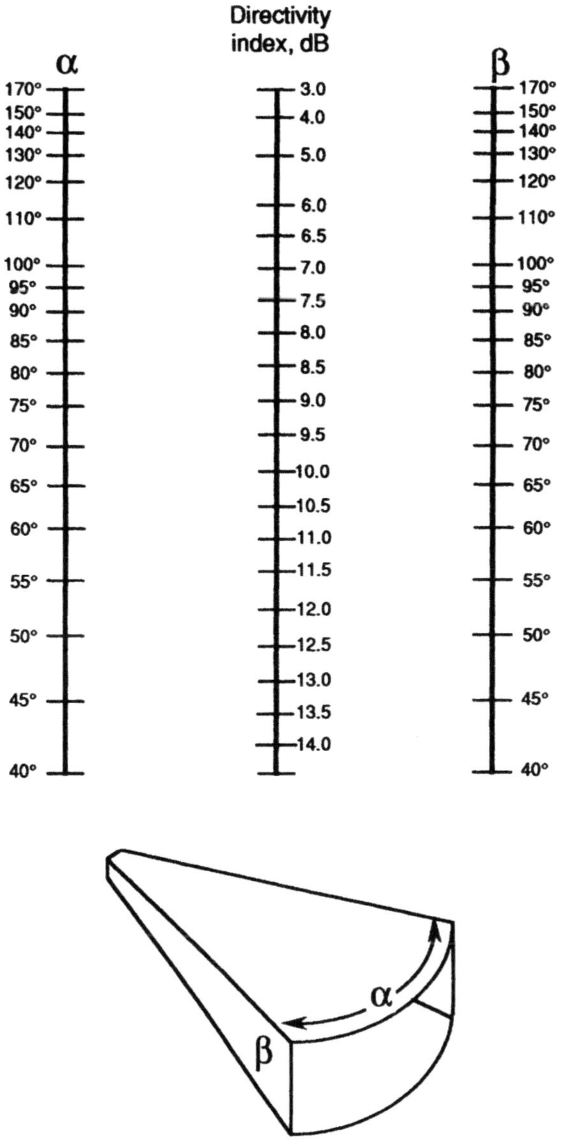

Figure 60. Directivity versus Horizontal and Vertical Beamwidth

61. Beamwidth and Directivity Characteristics of a Pair of 250-mm (10-in) Low-Frequency Transducers

For many applications in general sound reinforcement, low-frequency transducers are used in pairs to attain directivity response that is narrow in one plane and wide in the other. The beamwidth will, of course, be narrow in the plane in which the transducers are in line and wide in the perpendicular plane. The transducers are assumed to be in line in the vertical plane.

The data given here are taken as an average of a number of dual low-frequency systems built by several manufacturers. The mounting distance between the transducers is generally about one-fifth the diameter of a single transducer.

The data are presented up to a frequency of 1600 Hz, which is the useful upper frequency limit of such a system. Note that the DI approaches an asymtote of 3 dB at low frequencies, reflecting nominal 2π mounting conditions.

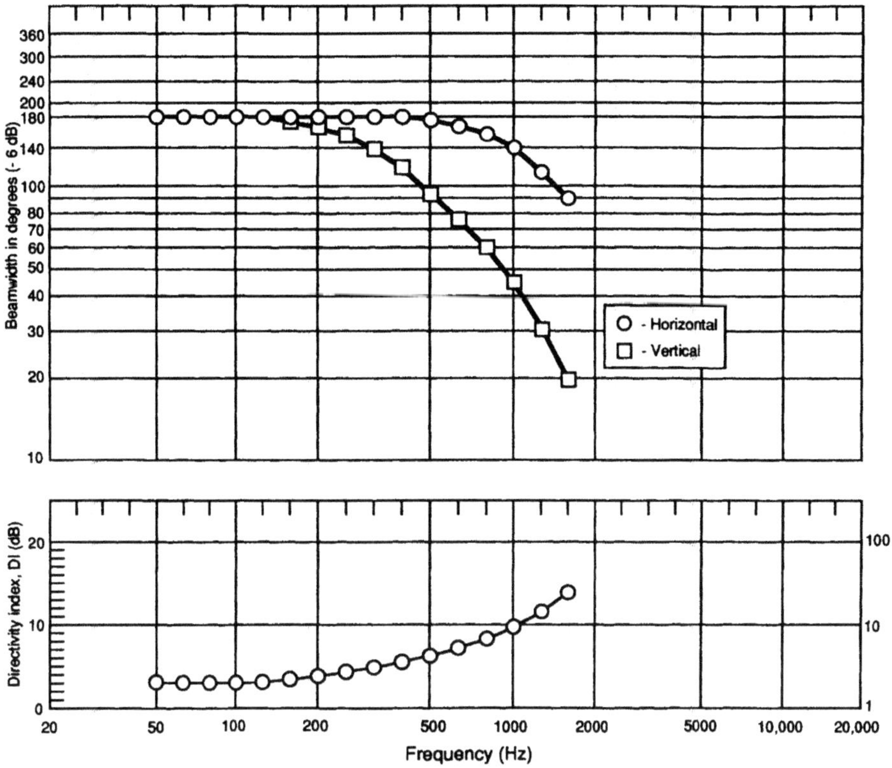

Figure 61. Beamwidth and Directivity Characteristics of a Pair of 250-mm (10-in) Low-Frequency Transducers

62. Beamwidth and Directivity Characteristics of a Pair of 300-mm (12-in) Low-Frequency Transducers

The data shown in Figure 62 are for a low-frequency system that is scaled upward in size from the one shown in section 61. The scale factor is approximately one-third octave. Again, 2π mounting conditions are assumed.

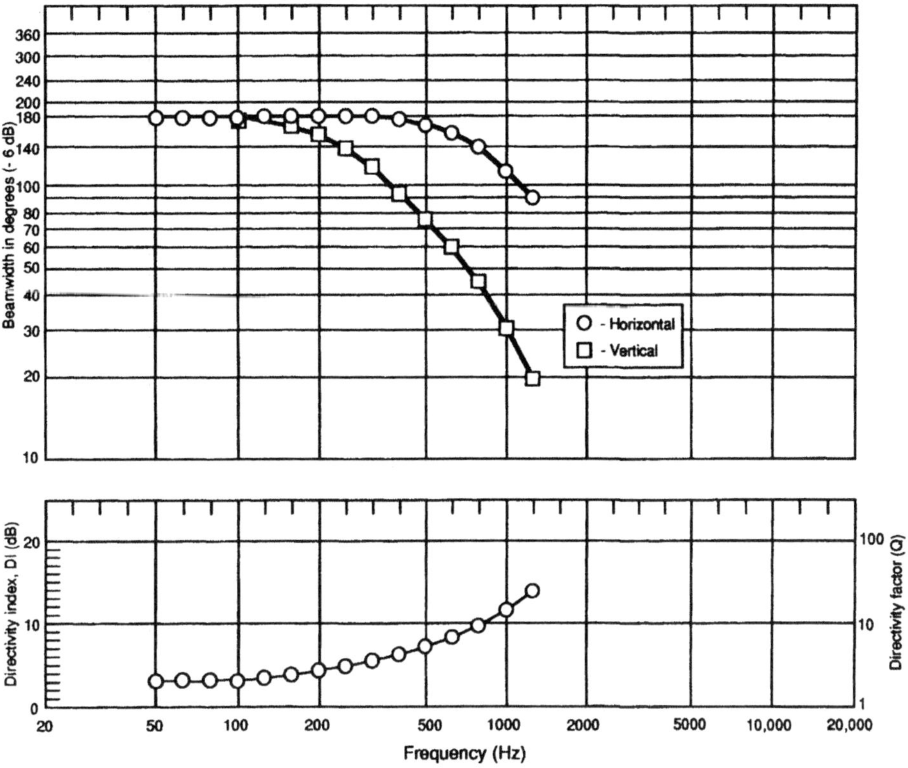

Figure 62. Beamwidth and Directivity Characteristics of a Pair of 300-mm (12-in) Low-Frequency Transducers

63. Beamwidth and Directivity Characteristics of a Pair of 380-mm (15-in) Low-Frequency Transducers

The data shown in Figure 63 are for a low-frequency system that is scaled upward in size from the one shown in section 62. The scale factor is approximately one-third octave.

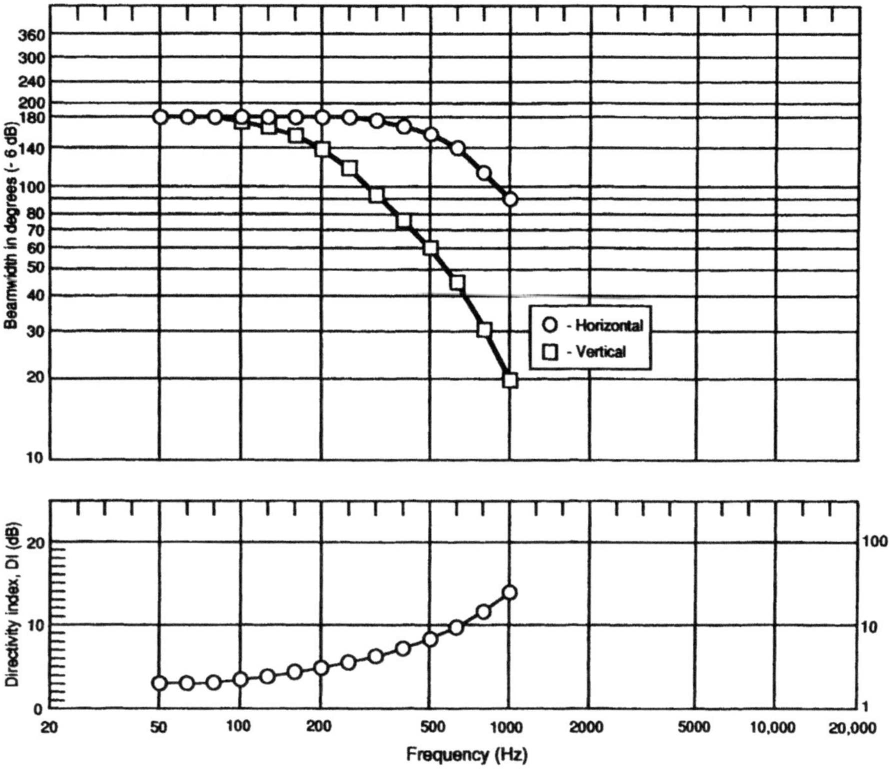

Figure 63. Beamwidth and Directivity Characteristics of a Pair of 380-mm (15-in) Low-Frequency Transducers

64. DISTRIBUTED LOUDSPEAKER LAYOUT: HEXAGONAL ARRAY

Distributed loudspeaker systems are normally mounted in enclosures that are located in suspended ceiling sections. We strongly recommend that only those loudspeakers with −6 dB beamwidth control of 90° up to 2 kHz be used. The analysis given here assumes that to be the case.

Three mounting densities are discussed in this section. Figure 64A shows section and projected ceiling views of what is called an "edge-to-edge" configuration. The circles indicate the isobar corresponding to the −6 dB contour of the loudspeaker.

It is obvious by inspection that the adjacent distance between loudspeakers, X, is equal to 2L, the distance between the ceiling and the listener's ear level. For seated listeners, values of 1 meter (4 feet) are normally used; for standing listeners, values of 1.8 meters (6 feet) are normally used.

The configuration shown in Figure 64B is known as "minimum overlap." In this case, the value of X is equal to 1.73 L.

The configuration shown in Figure 64C is known as "center-to-center." The value of X is equal to L.

The maximum variation in level of the three configurations is shown in Figure 64D. These data were calculated for direct field effects only. The approximate number of loudspeakers required for a given installation is given by the following equation:

$$\text{Quantity} = \text{Area}/.866X^2 \quad (64.1)$$

where area and X are expressed in consistent units.

EXAMPLE: A meeting room has dimensions of 20 × 35 × 4 meters. Determine the number of ceiling loudspeakers that will be necessary to cover the audience seating area with a hexagonal loudspeaker array with center-to-center coverage.

First, we determine that L is equal to 3 meters (4 − 1). The number of loudspeakers can then be calculated as follows:

$$\text{Quantity} = (20 \times 35)/(.866)3^2 = 90 \text{ loudspeakers}$$

References: Davis and Davis (1987)
Eargle (1989)
Foreman (1987)

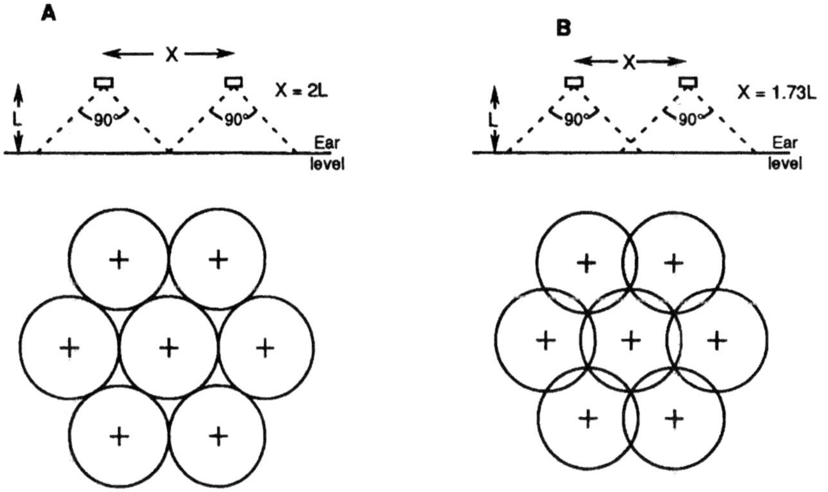

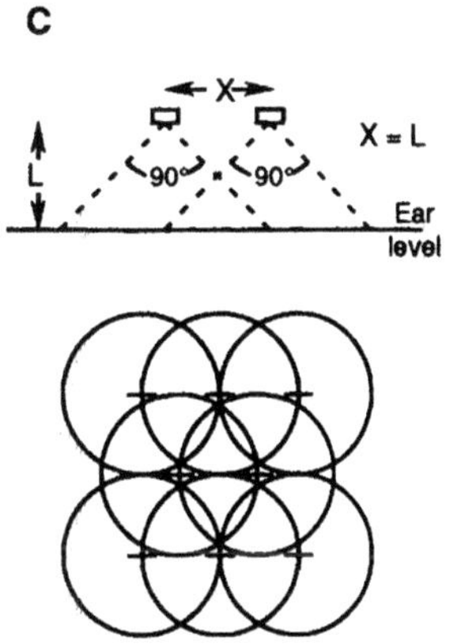

D

Configuration:	Level variation:
Edge-to-edge (A)	5.4 dB
Minimum overlap (B)	2.6 dB
Center-to-center (C)	1.2 dB

Figure 64. Distributed Loudspeaker Layout: Hexagonal Array

65. Distributed Loudspeaker Layout: Square Array

An alternate approach for laying out distributed systems is discussed in this section, and the same 90° coverage angle at 2 kHz is assumed for the loudspeakers.

The edge-to-edge array is shown in Figure 65A. Here, X is equal to 2L. Minimum overlap is shown in Figure 65B; X is equal to 1.5L. For the center-to-center configuration (Figure 65C), X is equal to L.

The maximum variation in level of the three configurations is shown in Figure 65C.

For all configurations the number of loudspeakers required is:

$$\text{Quantity} = \text{Area}/X^2 \quad (65.1)$$

where area and X are expressed in consistent units.

EXAMPLE: A meeting room has dimensions of 20 × 35 × 4 meters. Determine the number of ceiling loudspeakers that will be necessary to cover the audience seating area with a square loudspeaker array with center-to-center coverage.

First, determine that L is equal to 3 meters (4 − 1). The number of loudspeakers can then be calculated as follows:

$$\text{Quantity} = (20 \times 35)/3^2 = 78 \text{ loudspeakers}$$

A comparison with section 64 shows that the hexagonal array, with its requirement of 90 loudspeakers, affords an overall level variation of 1.2 dB, as opposed to 1.8 dB with the square array and its lesser quantity of loudspeakers.

References: Davis and Davis (1987)
Eargle (1989)
Foreman (1987)

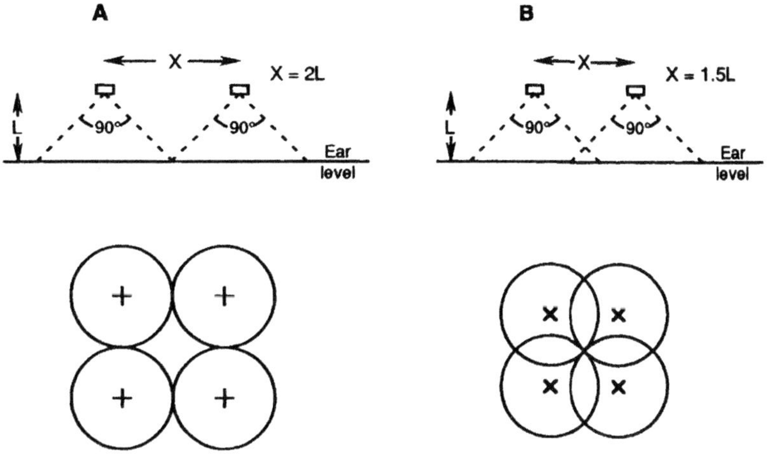

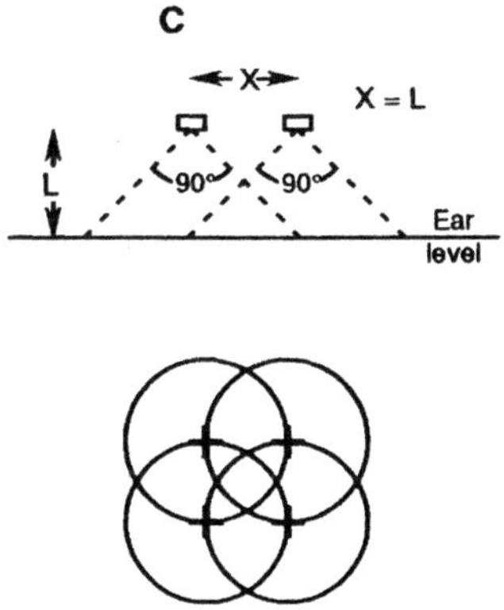

D

Configuration:	Level variation:
Edge-to-edge (A)	4.35 dB
Minimum overlap (B)	2.0 dB
Center-to-center (C)	1.4 dB

Figure 65. Distributed Loudspeaker Layout: Square Array

66. Dividing Networks: 6 dB per Octave Slopes

The data given in Figure 66 are based on the following equations:

$$C = 1/(2\pi f X_C) \quad \ldots\ldots (66.1)$$

$$L = X_L/2\pi f \quad \ldots\ldots (66.2)$$

In these equations, X_C and X_L represent the nominal impedances of the transducer loads at the chosen crossover frequency. Normally, these values will be 4 or 8 ohms. The crossover frequency is represented by f, and C and L are the calculated values of capacitance and inductance.

In practice, individual transducers have impedances that vary with frequency. Values of capacitance and inductance given in Figure 66A may need to be adjusted to a greater or lesser degree in order to attain the desired response. All values taken from the graph must actually be connected to the target transducer and measured to ensure that the desired response is obtained.

EXAMPLE: Find the values of capacitance and inductance for a 6-dB-per-octave dividing network for 8-ohm transducers crossing over at 500 Hz.

Locate 500 Hz on the abscissa and read upward to the bold line. Then read the value of 40 microfarads on the left ordinate. The value of inductance is read as 2.5 millihenrys on the right ordinate. Values of both capacitance and inductance may be rounded off to the nearest stock component values. The network topology is shown in Figure 66B.

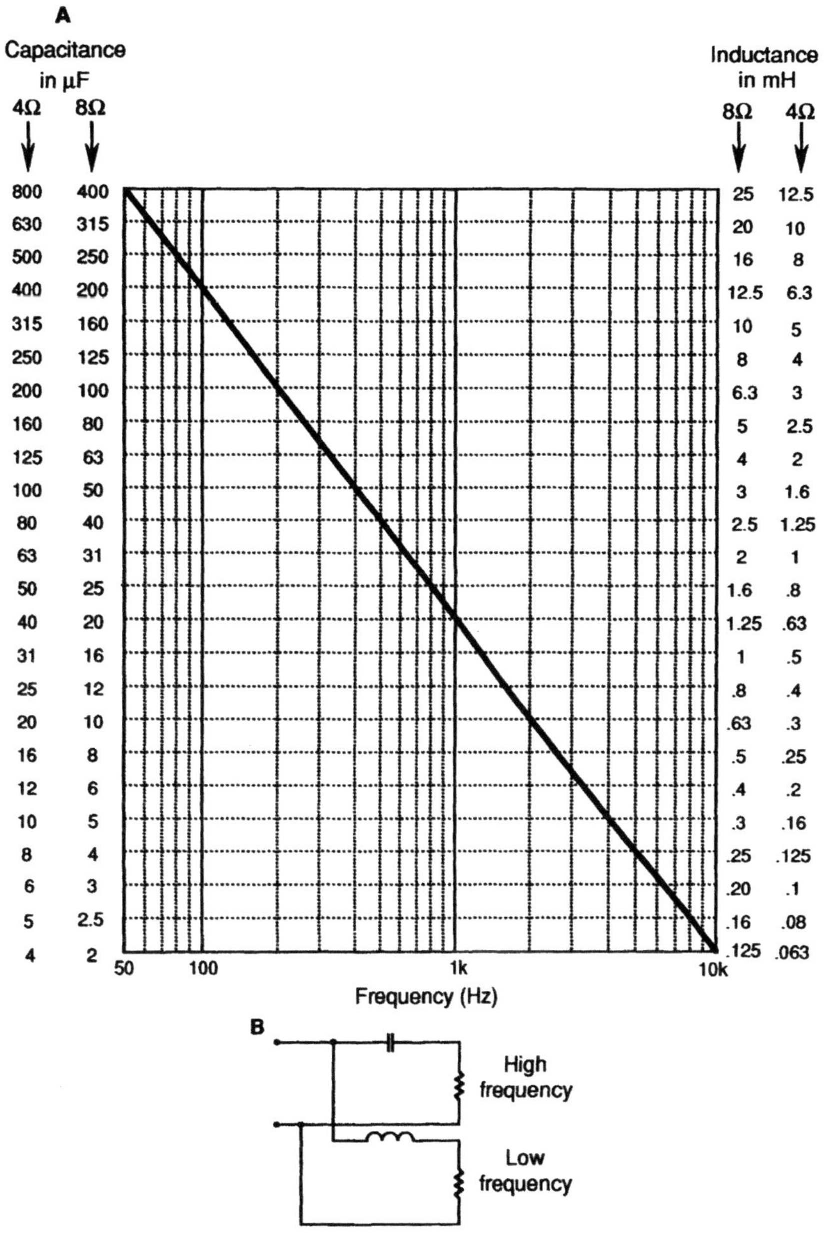

Figure 66. Dividing Networks: 6 dB per Octave Slopes

67. Dividing Networks: 12 dB per Octave Slopes

The data given in Figure 67 are based on the following equations:

$$C = 0.707/(2\pi f X_C) \quad (67.1)$$

$$L = 1.414X_L/2\pi f \quad (67.2)$$

In these equations, X_C and X_L represent the nominal impedances of the transducer loads at the chosen crossover frequency. Normally, these values will be 4 or 8 ohms. The crossover frequency is represented by f, and C and L are the calculated values of capacitance and inductance.

In practice, individual transducers have impedances that vary with frequency. Values of capacitance and inductance given in Figure 67A may need to be adjusted to a greater or lesser degree in order to attain the desired response.

EXAMPLE: Find the values of capacitance and inductance for a 12 dB per octave dividing network for 4-ohm transducers crossing over at 1 kHz.

Locate 1 kHz on the abscissa and read upward to the bold line. Then read the values of 28 microfarads on the left ordinate and 0.9 millihenrys on the right ordinate. Values of both capacitance and inductance may be rounded off to the nearest stock components values. The network topology is shown in Figure 67B.

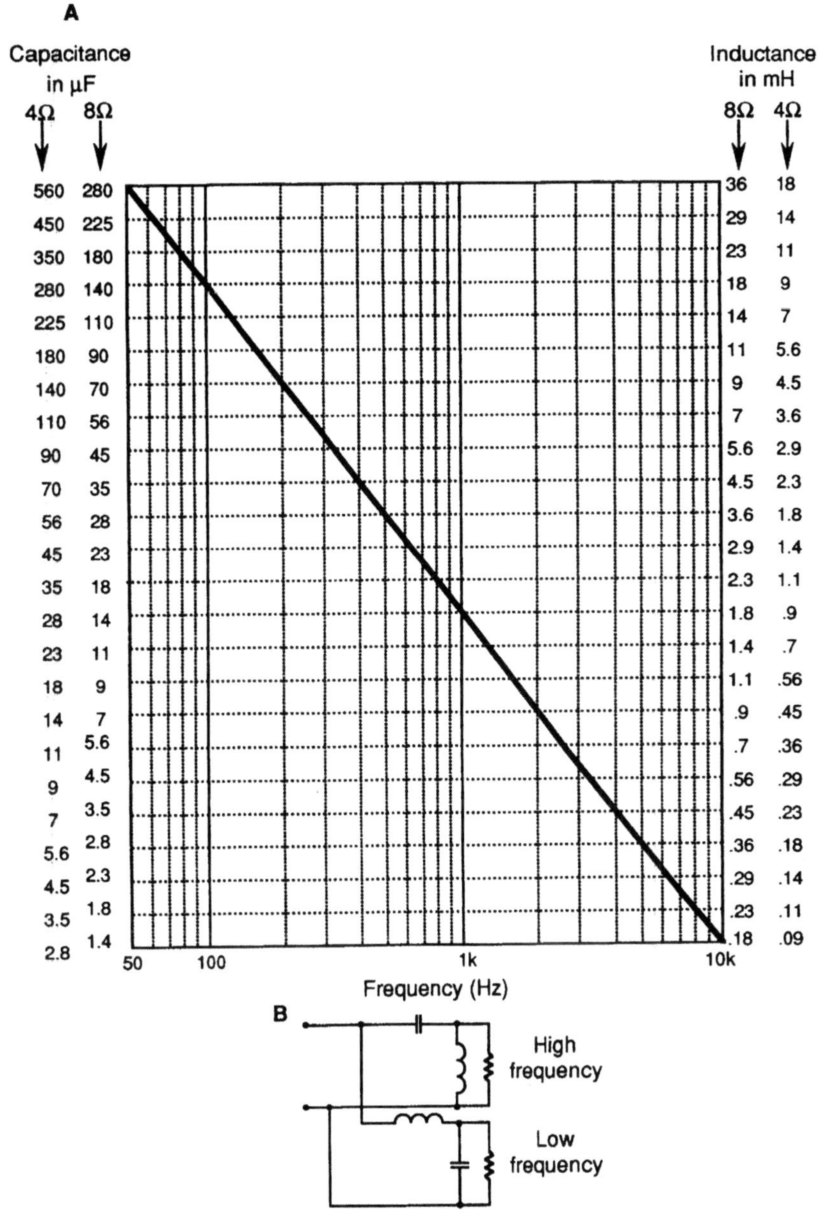

Figure 67. Dividing Networks: 12 dB per Octave Slopes

68. Porting Data for Vented Loudspeaker Enclosures

Figure 68 presents information on determining vented port dimensions that will ensure a desired resonance frequency for an enclosure of a specific volume.

EXAMPLE: For a 225-liter enclosure tuned to 40 Hz, determine the choices of port diameter and port length that may be used.

At Figure 68A, locate the value of 225 liters. With a straightedge, connect that point with the value of 40 Hz at Figure 68B. Extend that line so that it intersects with the construction line at Figure 68C. At the point of intersection with line C, scribe a horizontal line into the graph.

The various options include the intersections of the curves and the vertical dashed lines. For example, a port length of 10 cm intersects the curve for a port diameter of 200 mm; another possibility is a port length of 4 cm and a port diameter of 160 mm.

Reference: Small (1973)

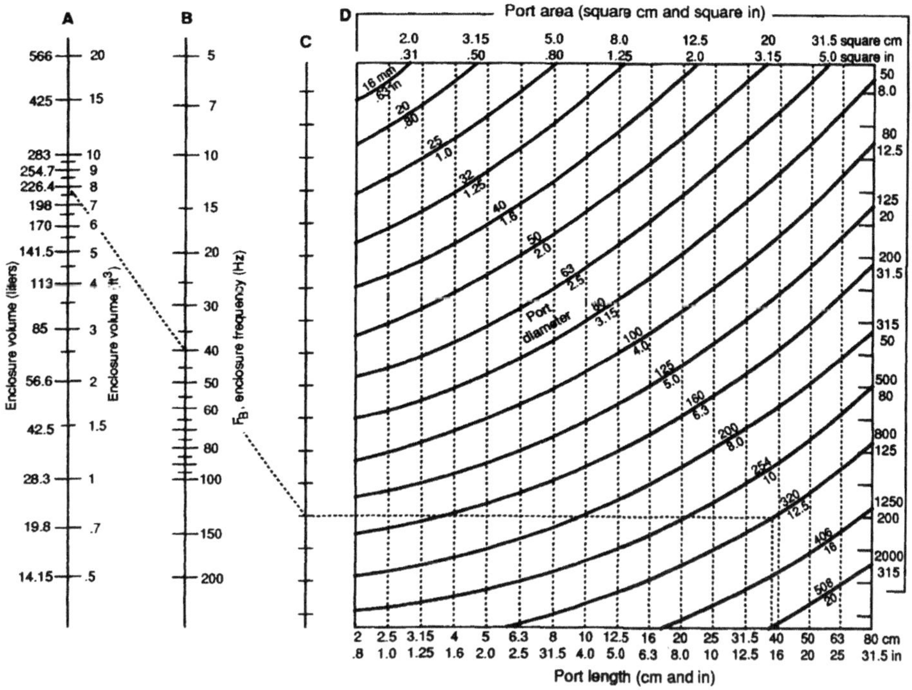

Figure 68. Porting Data for Vented Loudspeaker Enclosures

69. THIELE-SMALL PARAMETERS FOR LOW-FREQUENCY HORN APPLICATIONS

It is customary to use cone transducers as drivers for low-frequency horns. The pertinent Thiele-Small parameters give an excellent idea of how well the horn will perform.

As seen in Figure 69, there will be a region of flat power response bounded by f_{LC} at the low end and f_{HM} at higher frequencies. Beyond f_{HM}, there will be two additional breakpoints in the response envelope due to voice coil inductance and the effect of placing the air chamber directly in front of the transducer. Each breakpoint will add a 6 dB/octave rolloff to the overall response curve. The various breakpoint frequencies are defined as follows:

$$f_{LC} = (Q_{ts})f_s/2 \qquad (69.1)$$

$$f_{HM} = 2(f_s)/Q_{ts} \qquad (69.2)$$

$$f_{HVC} = R_e/\pi L_e \qquad (69.3)$$

$$f_{HC} = (2Q_{ts})f_s(V_{as}/V_{fc}) \qquad (69.4)$$

where: Q_{ts} = total Q of the transducer
f_s = free-air resonance of the transducer
R_e = voice coil dc resistance (in ohms)
L_e = voice coil inductance of the transducer (in henrys)
V_{as} = volume of air that provides a restoring force equal to that of the transducer's mechanical compliance (in liters)
V_{fc} = volume of the front air chamber (in liters)

Reference: Keele (1977)

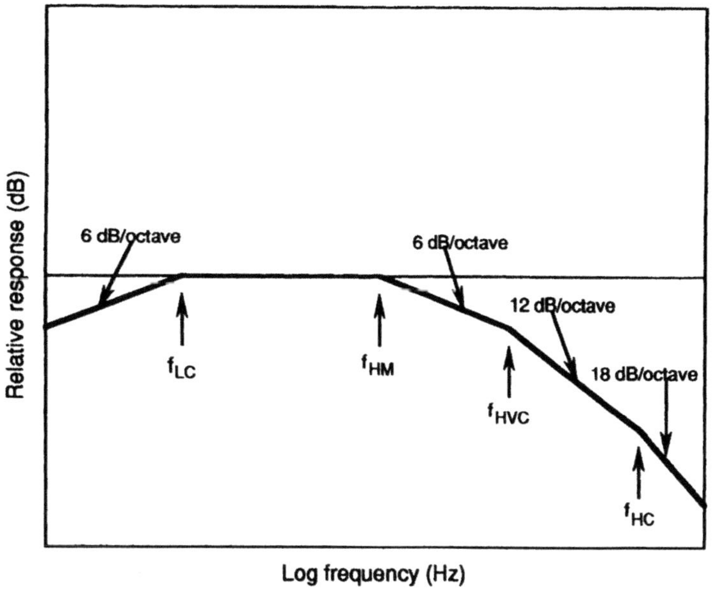

Figure 69. Thiele-Small Parameters for Low-Frequency Horn Applications

70. SIMPLE LINE ARRAYS

Simple line arrays are defined as those whose components are all driven by the same signal and whose directivity, taken individually, is negligible. Such devices have been used over the years for certain speech reinforcement applications where tight pattern control and high acoustical power output are not required. The ubiquitous sound column fits this general description.

The response in the far field of a linear array, as measured in the plane of the array, is given by the following equation:

$$R(\theta) = \frac{\sin [\frac{1}{2} Nkd \sin \theta]}{N \sin [\frac{1}{2} kd \sin \theta]} \quad \text{(70.1)}$$

where N is the number of elements in the array, k is equal to $2\pi f/c$, d is the spacing of the elements in the array, and θ is the measurement angle relative to the normal to the array. The speed of sound is denoted by c. The quantities in brackets must be evaluated in radians.

Figure 70A–D shows the theoretical response of a four-element line array ($N = 4$), with spacing between the components of 0.2 meter, for frequencies of 200, 350, 500, and 1000 Hz. It should be understood that the radiation patterns shown here are uniform in the horizontal plane, assuming that the line array is oriented vertically. We further assume that these measurements are made at a distance from the array at least equal to 10 times the length of the array.

Note that the directivity of the array is clearly a function of frequency, as can be seen in the polar graphs. Further indication of this is shown in the plots of directivity factor for various values of N as seen in Figure 70E. For all values of N, there is a peak in the directivity factor very close to the value of $d/\lambda = 1$, diminishing quickly above that value and then increasing gradually.

There are some important tradeoffs in design: At higher frequencies, line arrays with more elements will generally produce side lobes that are no greater than 10 dB below the main lobe, whereas those composed of fewer elements may produce very strong side lobes. On the other hand, the larger arrays will exhibit a greater overall variation of directivity factor over the useful range than will those composed of fewer elements.

These particular problems can be addressed through electrical or acoustical tapering, or shaping, of the frequency response of the various elements, enabling them to maintain more uniform directivity with respect to frequency over a broad frequency range.

References: Klepper and Steele (1963)
Kuttruff (1979)

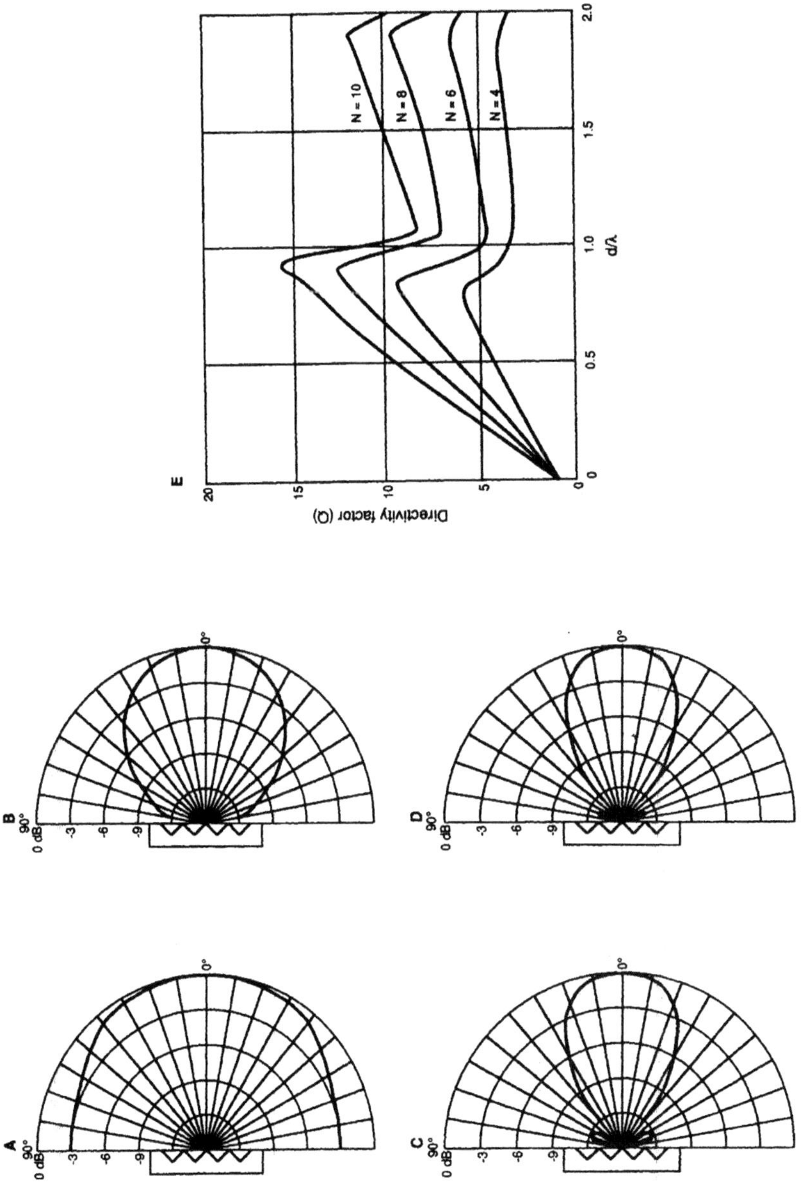

Figure 70. Simple Line Arrays

PART THREE

MICROPHONES

71. Nomograph for Microphone Output Power and Voltage versus Microphone Impedance

Engineers generally deal with microphone voltage sensitivity. The standard method is to state the unloaded output voltage when the microphone is placed in a sound field of 94 dB L_p. This is appropriate, since today's high-quality capacitor microphones are always operated into an essentially open circuit, one that is at least 10 times the nominal impedance of the microphone.

However, in much of the literature dealing with dynamic microphones, we find references to older microphone rating methods. The nomograph given in this section will be useful in relating microphone output voltage, output power, microphone impedance, and the load impedance.

To use the nomograph in Figure 71, place a straightedge so that a line may be drawn between the microphone's impedance, as read in C, and the microphone's power sensitivity, as read at B. When this is done, the microphone's output voltage can be read at A.

EXAMPLE: A dynamic microphone has an impedance of 200 ohms and a power sensitivity rating of −40 dB re 1 mW/10 dynes per cm^2. What will its open-circuit output voltage be if it is operated in a sound field of 10 dynes per cm^2?

Place a straightedge so that it intersects the 200 marker at C and the −40 dB marker at B. Then read +19 dB along the right scale of A. The actual output voltage level will be 19 dB relative to 1 mV, or:

Output voltage = $10^{19/20}$ = 9 millivolts

If this microphone were loaded with its matching 200 ohm impedance, the output voltage level would be +13 dB, as read at the left scale of A. Solving this gives an output voltage of 4.5 millivolts.

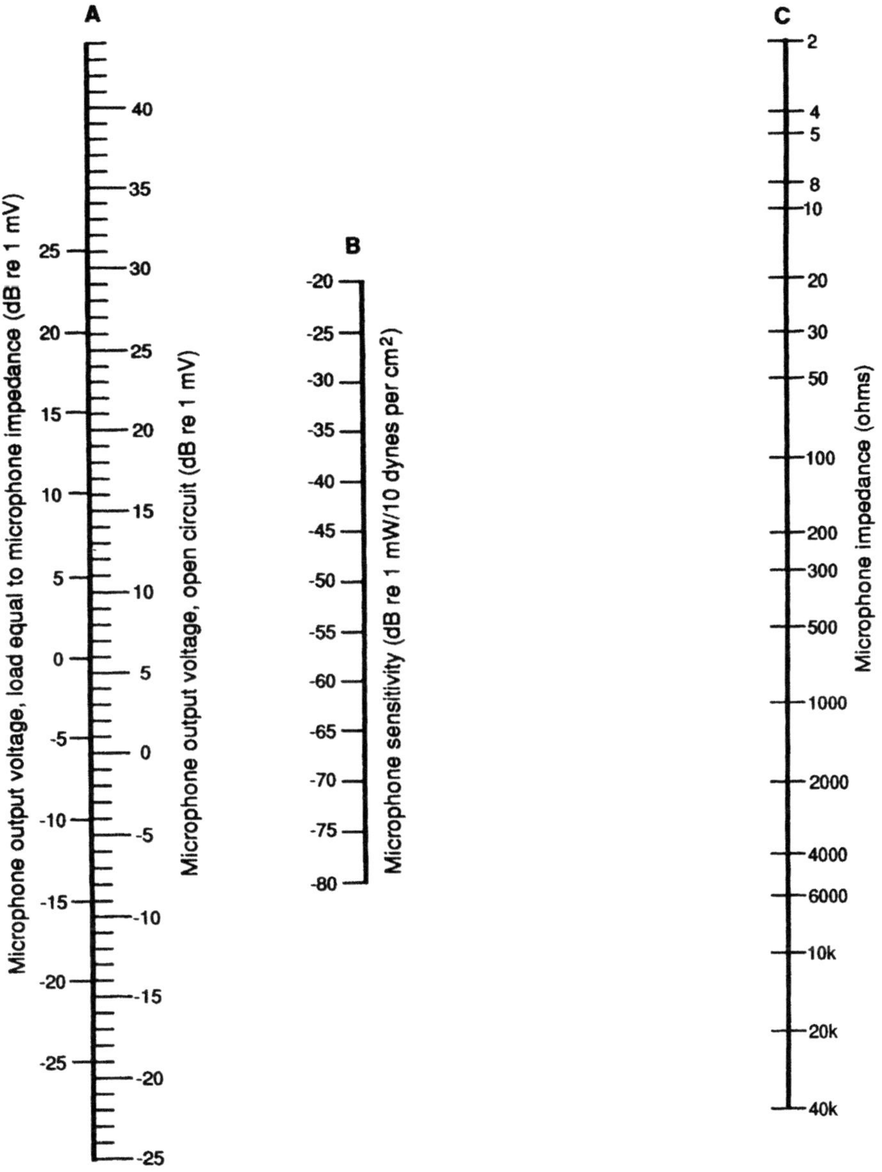

Figure 71. Nomograph for Microphone Output Power and Voltage versus Microphone Impedance

72. Microphone Self-Noise Rating Curves

Microphone self-noise is generally measured using the standard A-weighting curve, since this method provides a useful comparison between microphones that are used for music and speech recording played back at normal levels.

The A-weighting curve (see section 138) is often referred to by various standards document numbers in microphone specifications, but the simple designation dB(A) is invariably used as well.

Another common measurement standard is usually designated by CCIR 468. This curve is useful in comparing the relative self-noise of microphones whose outputs may be listened to at elevated levels. Both of these weighting curves are shown in Figure 72.

In practice, a microphone's noise level is measured by placing it in an absolutely quite environment and adding a constant correction factor dependent on the microphone's sensitivity to a standard metering system employing the desired weighting curve. For most studio-quality microphones, the difference between the A-weighting curve and the 468 curve will be about 11 dB. An examination of these two curves in the graph will show that there is approximately 11 dB between them in the 5–9 kHz range, where the high-frequency noise output of the microphones may be prominent.

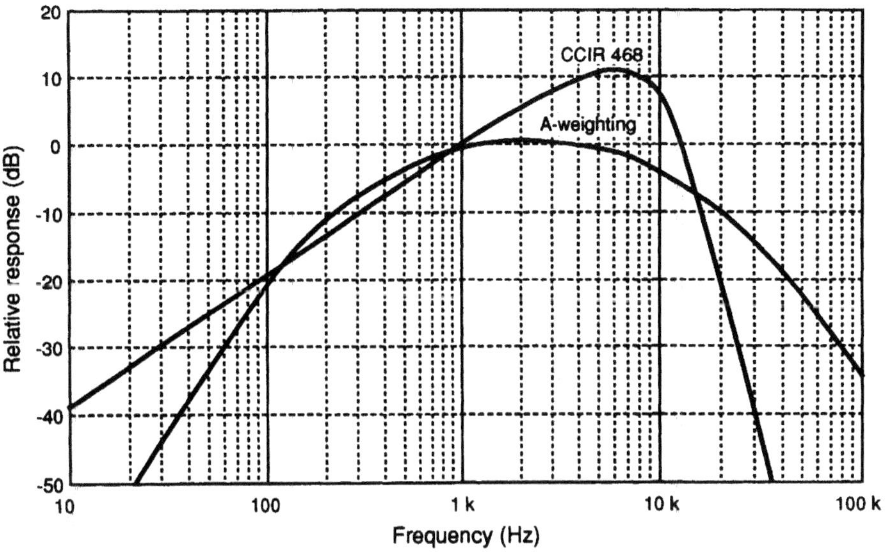

Figure 72. Microphone Self-Noise Rating Curves

73. EIA G_M Microphone Sensitivity Rating

The G_M microphone sensitivity rating dates from an earlier era when microphones looked into a load impedance equal to their source impedance and when microphone source impedances varied over a wide range. It was essentially a power output rating rather than a voltage output rating. It is presented here primarily for reference.

If we know both the unloaded output voltage of a microphone in a 74 dB L_p sound field and its source impedance, then the G_M value is given by the following equation:

$$G_M = S_V - 10 \log Z_n - 50 \quad (73.1)$$

where Z_n is the nominal microphone impedance as given at B in Figure 73 and S_V is as given at A

EXAMPLE: A microphone has an open circuit output voltage of 0.7 millivolts in a 74 dB L_p sound field. Its rated impedance is 200 ohms. Determine its G_M sensitivity rating.

The output voltage of 0.7 millivolts corresponds to S_V of −43 dB, and the impedance rating of 200 ohms falls in the 150-ohm range for Z_n. Noting that the decibel value in brackets for Z_n = 150 is approximately 22 dB, we can now complete equation 73.1:

$$G_M = -63 - 22 - 50 = -135 \text{ dB}$$

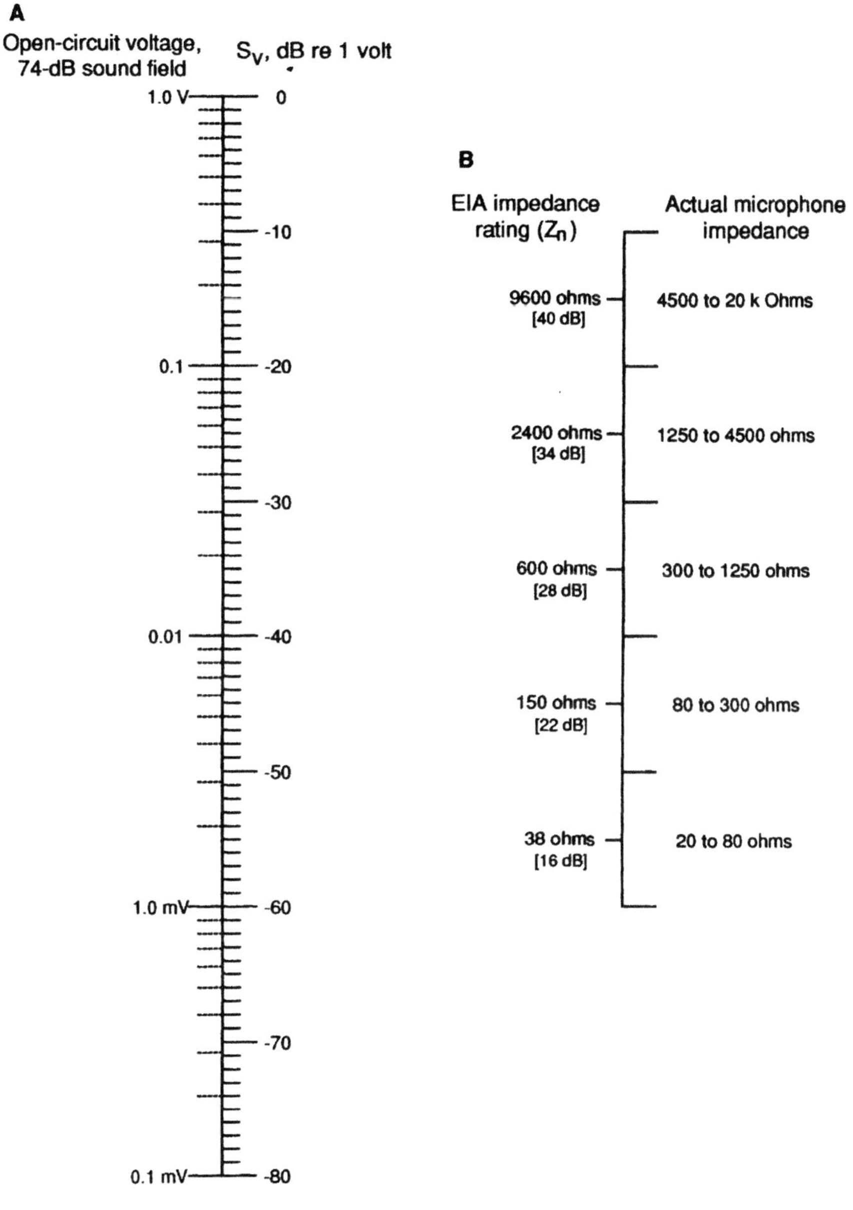

Figure 73. EIA G_M Microphone Sensitivity Rating

74. First-Order Microphone Pattern Data

The microphone patterns shown in Figure 74 are known as "first-order patterns" because their polar equations are composed of various ratios of unity and cos θ taken to the first power. The patterns range from omnidirectional, when the cosine term is zero, to bidirectional, when the constant term is zero. The various combinations of these two basic patterns yield the family of cardioid patterns.

The −3-dB and −6-dB pickup arcs are given as guides to selecting patterns for frontal coverage.

The response at 90° off-axis is given as guidance in selecting patterns for rejection of sound from the sides. Note that as the cardioids progress from subcardioid to hypercardioid, the 90° response ranges from −3 dB to −12 dB in approximate 3-dB steps.

The random energy efficiency (REE) is a single-number measurement of the on-axis directivity. REE is equivalent to 1/Q, where Q is the directivity factor and also a measure of the microphone's ability to discriminate against randomly arriving signals (as opposed to those arriving along the principal axis).

The distance factor (DSF) is a measure of the increase in working distance afforded by the cardioids, as compared with the omnidirectional pattern, for the same amount of rejection of randomly arriving signals. For example, for the same ratio of on-axis pickup to random pickup, a hypercardioid microphone may be used at twice the distance of an omnidirectional microphone.

SUMMARY OF FIRST-ORDER CARDIOID MICROPHONES

CHARACTERISTIC	OMNIDIRECTIONAL	BIDIRECTIONAL	SUBCARDIOID	CARDIOID	SUPERCARDIOID	HYPERCARDIOID
POLAR RESPONSE PATTERN	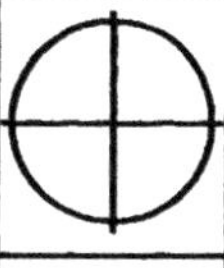	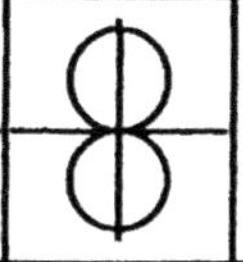		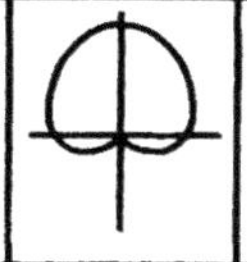		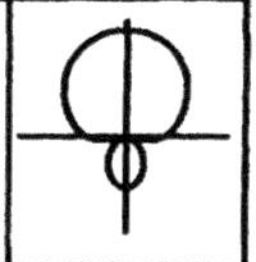
POLAR EQUATION	1	Cos θ	.7 + .3Cos θ	.5 + .5Cos θ	.37 + .63Cos θ	.25 + .75Cos θ
PICKUP ARC 3 dB DOWN	360°	90°	180°	131°	115°	105°
PICKUP ARC 6 dB DOWN	360°	120°	264°	180°	156°	141°
RELATIVE OUTPUT AT 90° (dB)	0	-∞	-3	-6	-8.6	-12
RELATIVE OUTPUT AT 180° (dB)	0	0	-8	-∞	-11.7	-6
ANGLE AT WHICH OUTPUT = ZERO	–	90°	–	180°	126°	110°
RANDOM ENERGY EFFICIENCY (REE)	1 0 dB	.333 -4.8 dB	.55 -2.5 dB	.333 -4.8 dB	.268 (1) -5.7 dB	.25 (2) -6 dB
DISTANCE FACTOR (DSF)	1	1.7	1.3	1.7	1.9	2

(1) MAXIMUM FRONT TO TOTAL RANDOM ENERGY EFFICIENCY FOR A FIRST-ORDER CARDIOID.
(2) MINIMUM RANDOM ENERGY EFFICIENCY FOR A FIRST-ORDER CARDIOID.

(Data presentation after Shure Brothers)

Figure 74. First-Order Microphone Pattern Data

75. MID-SIDE/XY CONVERSION DATA

Mid-side (MS) microphone arrays always consist of a side-responsive bidirectional microphone and a forward-responsive microphone of any first-order pattern. Through sum and difference matrixing, these two microphone outputs can be resolved to an equivalent left-right (XY) pair.

With the exception of a few simple cases, it is difficult to tell by inspection just what the MS equivalent for an arbitrary XY pair will be. The two graphs given here make the job simple.

EXAMPLE: For an XY microphone pair consisting of two normal cardioid patterns splayed at an angle of 120°, determine the corresponding MS pair.

Note first of all that 120° is equal to 2α. Find the cardioid symbol along the abscissa of Figure 75A. Then move upward until the 60° contour is intersected. Read to the ordinate at the right; the required M pattern will be given by a subcardioid microphone.

Then move on to Figure 75B to determine the required level for the S pattern. Locate the cardioid pattern along the abscissa and move upward to the intersection with the 60° curve; follow this curve to the ordinate at the right. Read the value of approximately +4.8 dB, indicating that the S component of the MS pair will have to be operated 4.8 dB higher than the M component. It is assumed that the M and S patterns have the same 0° sensitivity.

Reference: Hibbing (1989)

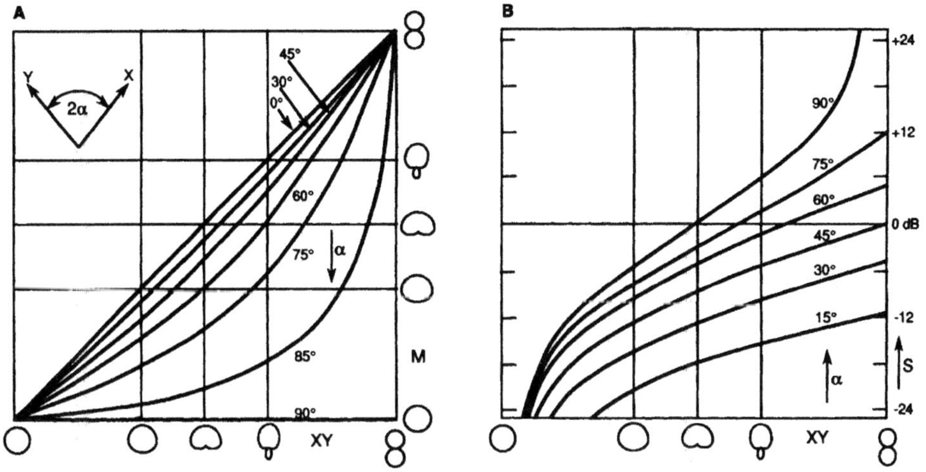

Figure 75. Mid-Side/XY Conversion Data

76. Random Energy Efficiency, Directivity Factor, and Distance Factor as a Function of Polar Pattern

The general polar equation of a first-order microphone pattern is given by:

$$\rho = A + B \cos \theta \quad \ldots \ldots \ldots \ldots \ldots \ldots \ldots \ldots \ldots \ldots (76.1)$$

where A + B = 1.

The Random energy efficiency (REE) of a first-order pattern is given by the following equation:

$$REE = 1 - 2B + 1.333B^2 \quad \ldots \ldots \ldots \ldots \ldots \ldots \ldots \ldots (76.1)$$

We can now plot the values of B from 0 to 1 and show the REE of the first-order patterns as a continuous function, as shown in Figure 76. Note that REE reaches a minimum value of 0.25 for the hypercardioid pattern.

The directivity factor (DF) is defined as:

$$DF = 1/REE \quad \ldots \ldots \ldots \ldots \ldots \ldots \ldots \ldots \ldots \ldots (76.3)$$

and this is plotted on the graph as a continuous function.

The distance factor (DSF) is defined as:

$$DSF = (DF)^{1/2} \quad \ldots \ldots \ldots \ldots \ldots \ldots \ldots \ldots \ldots \ldots (76.4)$$

and this is plotted on the graph as a continuous function.

These continuous functions can now be seen by inspection for all the cardioid patterns, as indicated along the abscissa.

REE, DF, and DSF are all measurements of a microphone's capability of responding preferentially to sound along the principal axis, as compared with the pickup of reverberant sound.

References: Bauer (1940)
Glover (1940)
Woram (1989)

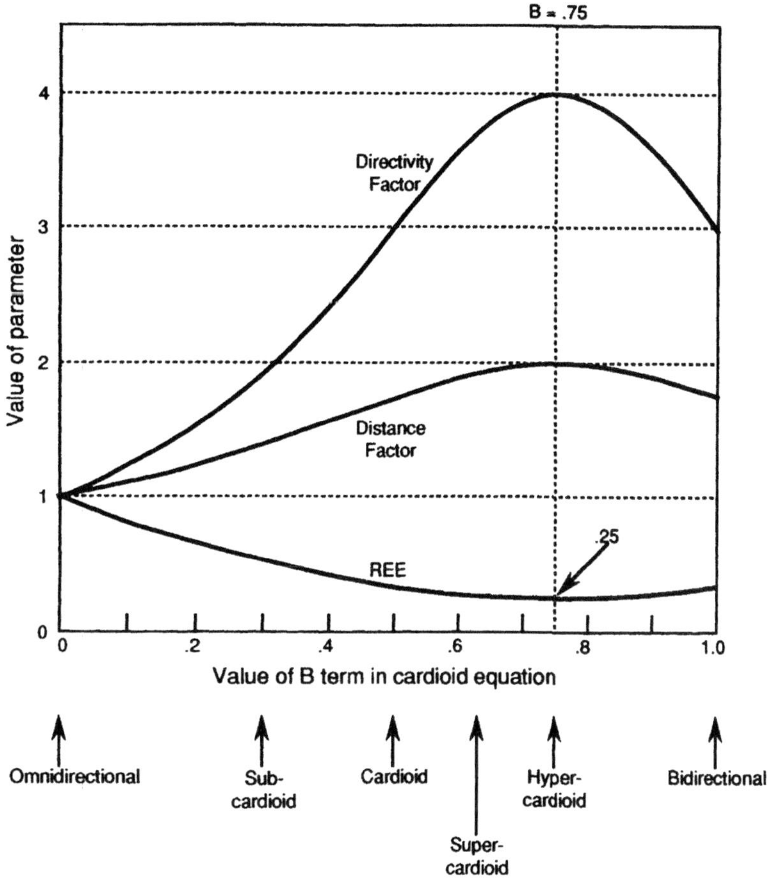

Figure 76. Random Energy Efficiency, Directivity Factor, and Distance Factor as a Function of Polar Pattern

77. Front-to-Total Ratio as a Function of Polar Pattern

The front-hemisphere random energy efficiency (REF) is given by the following equation:

$$REF = 0.5 - 0.5B + 0.167B^2 \qquad (77.1)$$

The front-to-total ratio (FTR) is then given by:

$$FTR = REF/REE \qquad (77.2)$$

This is plotted on the graph in Figure 77 as a continuous function. Note that FTR reaches a maximum value for the supercardioid pattern, indicating that this polar pattern maximizes the front hemisphere pickup, relative to pickup in all directions.

References: Bauer (1940)
Glover (1940)
Woram (1989)

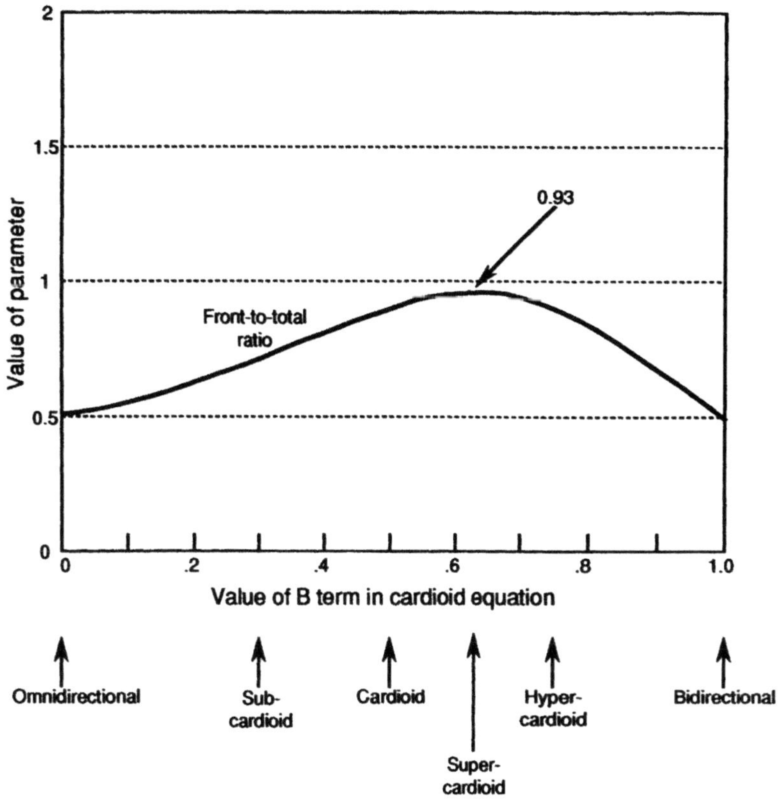

Figure 77. Front-to-Total Ratio as a Function of Polar Pattern

78. Front-Back Ratio versus Polar Pattern

The front-back ratio, often called the unidirectional index (UDI), is given by the following equation:

$$UDI = REF/REB \quad (78.1)$$

where:

$$REB = 0.5 - 1.5B + 1.167B^2 \quad (78.2)$$

This value is plotted on the graph in Figure 78 as a continuous function. Note that UDI reaches a maximum value for the supercardioid pattern.

References: Bauer (1940)
Glover (1940)
Woram (1989)

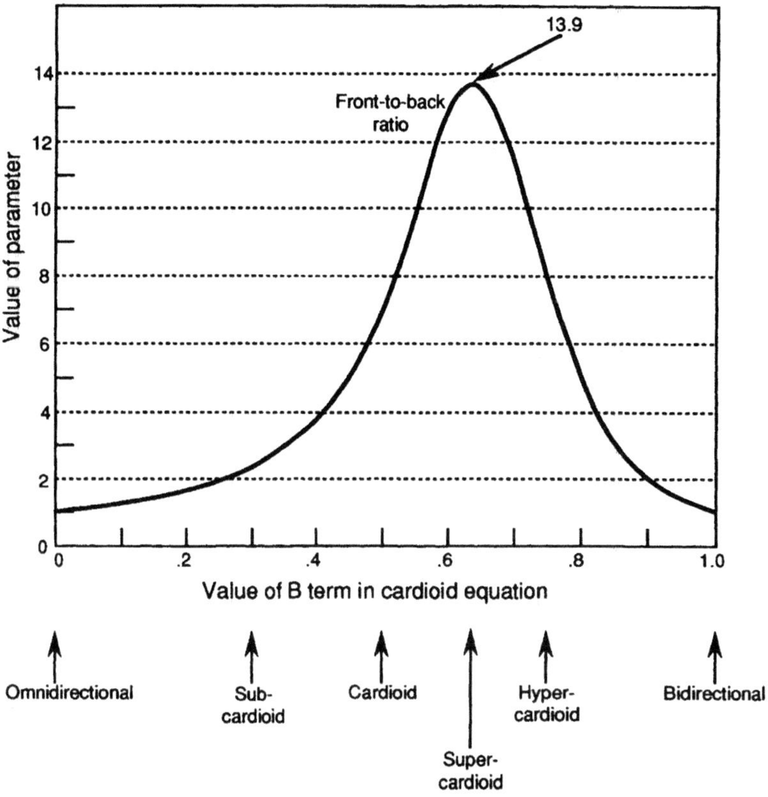

Figure 78. Front-Back Ratio versus Polar Pattern

79. Omni- and Bidirectional Components of the First-Order Cardioid Family

For graphic clarity, the relative components of omnidirectional and bidirectional patterns that are summed to produce cardioid patterns are shown in Figure 79.

Earlier variable-pattern microphones were often composed of separate omni- and bidirectional elements whose outputs were summed in the proportions required to produce the desired directional patterns.

References: Eargle (1981)
Eargle (1992)

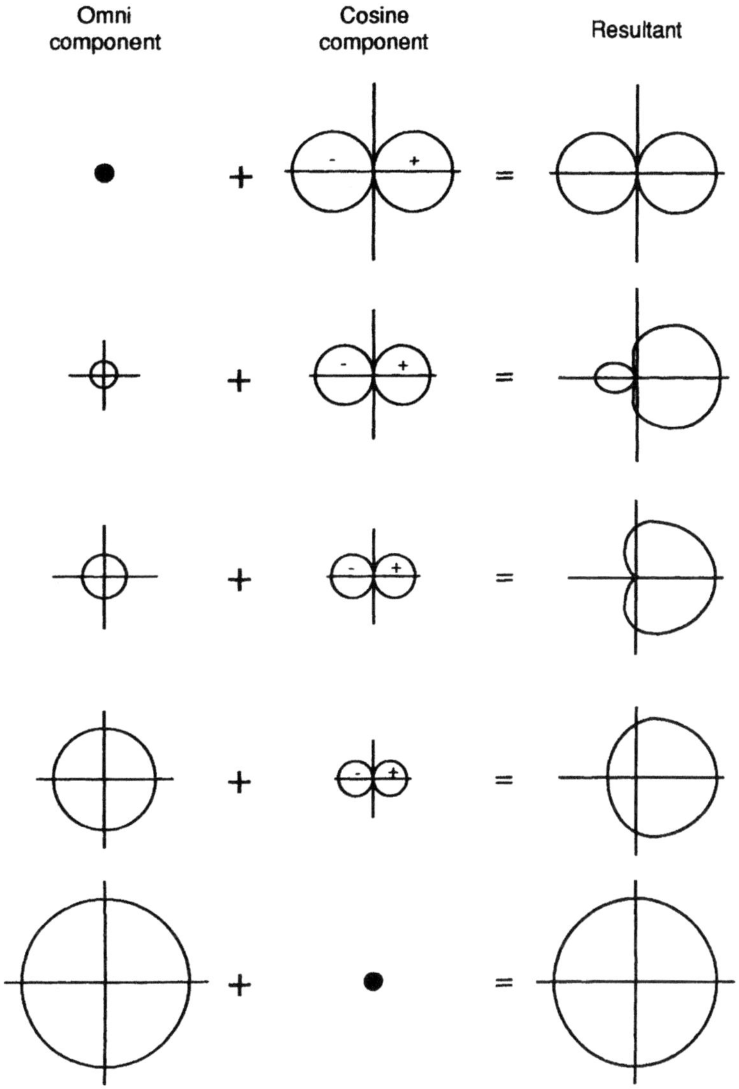

Figure 79. Omni- and Bidirectional Components of the First-Order Cardioid Family

80. Back-to-Back Cardioid Components of the First-Order Cardioid Family

Today, variable-pattern microphones almost universally make use of summing or subtracting back-to-back cardioid patterns to attain the desired response.

As an example, consider the first case shown in Figure 80—that of two equal and opposing cardioids subtracted from each other:

$$(0.5 + 0.5 \cos \theta) - (0.5 - 0.5 \cos \theta) = \cos \theta$$

In capacitor microphones of the Braunmühl-Weber design, the two cardioid patterns are nested in a unique design that has a single perforated backplate with a diaphragm on each side. The various summations are achieved by using a variable polarization voltage on one side and a fixed polarization voltage on the other.

Another capacitor design, also credited to Braunmühl and Weber, makes use of a single diaphragm with dual perforated backplates. Adjustable path lengths and acoustical damping around the assembly are used to produce the family of cardioids.

References: Eargle (1981)
Eargle (1992)
Sank (1985)

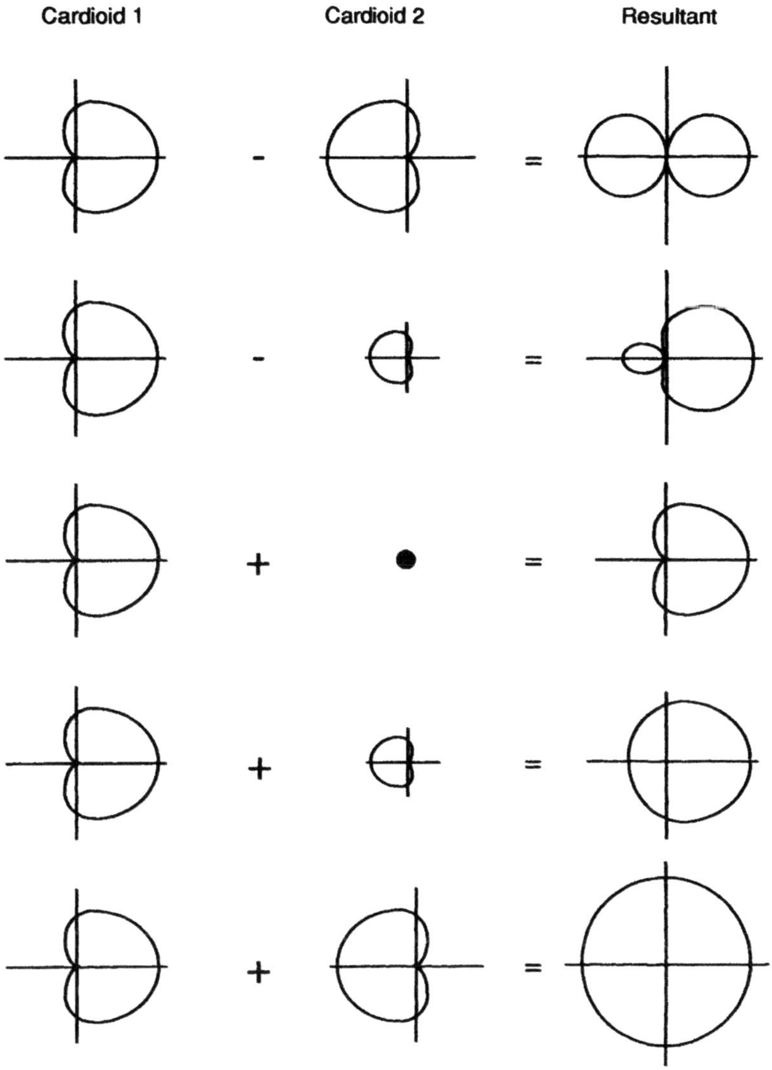

Figure 80. Back-to-Back Cardioid Components of the First-Order Cardioid Family

81. Splay Angles and Separation for Various Near-Coincident Stereo Microphone Arrays

Near-coincident-stereo microphone arrays rely primarily on intensity cues for spatial imaging in the lateral plane. At frequencies above about 1 kHz, however, these arrays may convey a certain diffuseness, due to their spacing and the fact that time relationships at high frequencies become significant.

There is little that is truly unique about any specific microphone pattern, splay angle, or separation in defining a near-coincident array; we might visualize a continuum of values in pattern, spacing, and angular orientation. The four examples shown in Figure 81 have been documented and have gained a degree of acceptance.

The ORTF (Office de Radiodiffusion-Télévision Français) array was developed by the French national broadcasting group and is widely used.

The NOS (Nederlandsche Omroep Stichting) array was developed by the Dutch broadcasting group. With its slightly wider spacing than in the ORTF array, there are stronger time cues for off-axis sound sources; these are partially offset by a narrower splay angle.

The Faulkner array, with its 0° splay angle, seems to rely entirely on time cues in conveying lateral localization, except for close-in sources where parallax comes into play.

With its very close spacing, the Stereo 180 array seems to barely qualify as near-coincident. Its principal characteristic is the contribution of opposite polarity relationship between stereo channels for sounds originating at the angular extremes of the array. These are, of course, caused by the fact that the back lobes of the hypercardioid patterns are of opposite polarity to the primary lobes. Under ideal monitoring conditions, these polarity relationships can give rise to images outside the loudspeaker base.

References: Eargle (1992)
Ceoen (1972)
Faulkner (1981)
Olson (1979)

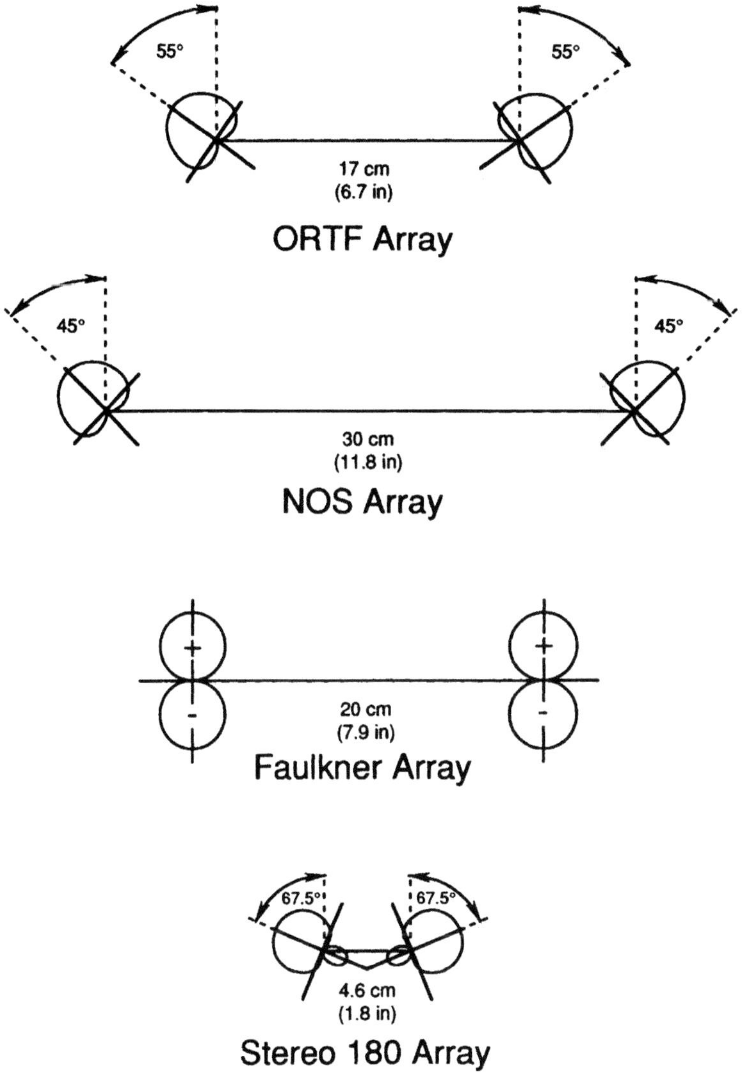

Figure 81. Splay Angles and Separation for Various Near-Coincident Stereo Microphone Arrays

82. Mid-Side (MS) and XY Microphone Pairs

Several MS microphone pairs are given in Figure 82, along with their resolution into stereo form, or XY. The relevant equations for this are:

$$X = 0.5(M + S) \quad (82.1)$$

$$Y = 0.5(M - S) \quad (82.2)$$

The form shown at Figure 82A is the earliest example of MS; it was described by Blumlein in 1931. It is of limited use inasmuch as its basic pickup is for two sources facing one another.

The form shown at Figure 82B is a popular one. It has the characteristic of picking up sound intensity with uniformity from all directions in the horizontal plane. Given the choice of using a pair of dipole microphones in the XY mode or via MS, many recording engineers will opt for the latter, since it exhibits better high-frequency stability along the median plane.

The form shown at Figure 82C is the more or less standard form of MS in which the M component is a cardioid microphone. The XY resolution is approximately equivalent to a pair of supercardioid microphones played at an angle of ±45°.

As we saw in section 75, there is a continuum of MS microphone forms, both in terms of M pattern and the ratio of M to S. Each one of these forms will have a unique XY resolution.

References: Blumlein (1931)
Dooley and Streicher (1985)
Eargle (1992)

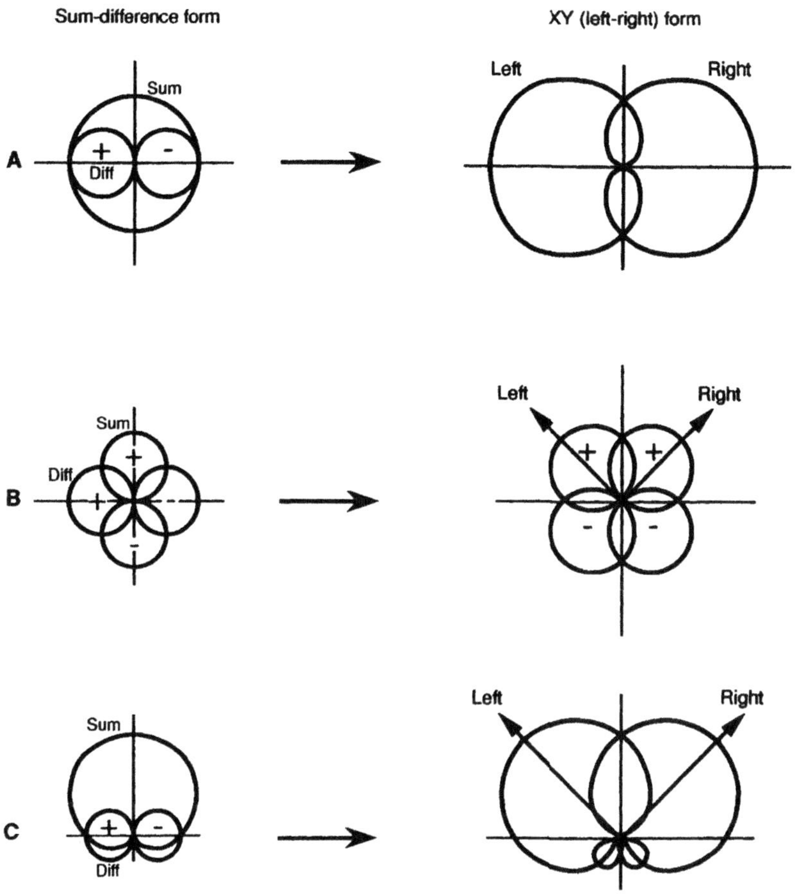

Figure 82. Mid-Side (MS) and XY Microphone Pairs

83. Multipath and Multimicrophone Interference Effects

When a single microphone receives both direct and reflected signals from the same source, there will be some degree of interference between the two signals. A similar situation occurs when two spaced microphones receive delayed signals that are later added together.

If the signal is of variable frequency, the interference pattern will consist of alternating peaks and dips in response, as a function of frequency. This is usually referred to as "combing" or "comb filtering," because the resulting response, plotted against frequency, resembles the tines of a comb.

The path length difference between the two signals can be converted into an equivalent delay time, D (sec):

$$D = \text{distance (meters)}/340 \quad \text{(83.1)}$$

The frequencies at which signal reinforcements will take place is given by the following equation:

$$f = n/D \quad \text{(83.2)}$$

where n is equal to integer values (1, 2, 3, . . . , n).

Cancellations will take place at frequencies that are spaced midway between the reinforcement frequencies. These are given by:

$$f = (2n + 1)/2D \quad \text{(83.3)}$$

Where n is defined as before. This equation results in the frequency series: 3/2, 5/2, 7/2, etc.

If the interfering signals are equal in amplitude, then the sum will be equal to two and the difference to zero. If they are of a different ratio, then less extreme sum and difference values will result.

The curves given in Figure 83 show the variation in response in both amplitude and signal level for five values of signal ratios.

Obviously, the use of directional microphones in reflection-prone environments will minimize this kind of interference.

The data presented here can also be used to determine the spacing of peaks and dips along the frequency axis resulting from signals that are delayed electrically. In this case, the value of D is simply the electrical delay.

Reference: Eargle (1992)

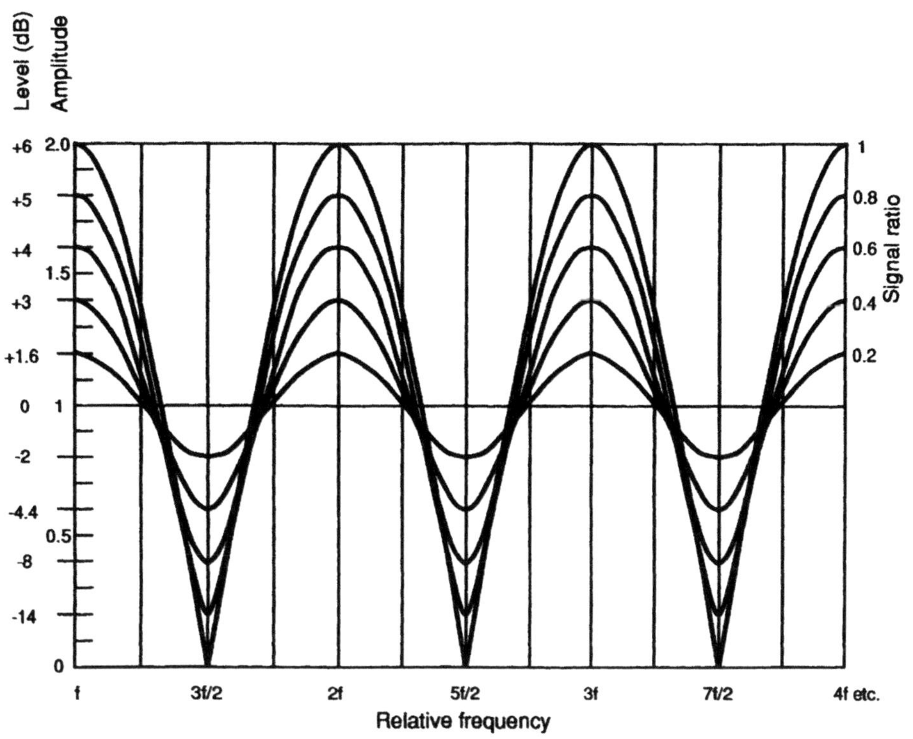

Figure 83. Multipath and Multimicrophone Interference Effects

84. Effect of Dipole Dimension on Directional Microphone Frequency Response

All directional microphone patterns can be analyzed as combinations of omnidirectional and dipole (figure 8) patterns. The finite dimensions of the dipole element determine the half wave-length at which signals entering the front of the dipole will be cancelled by the same signal entering the rear of the dipole. This is indicated in Figure 84A and Figure 84B as $2f_T$, at this frequency, there will be a signal cancellation at the dipole element. The useful upper limit of the dipole is an octave below this, or f_T.

Since dipole response is proportional to pressure gradient, or the difference between the pressures at the two openings of the dipole, it has a characteristic 6 dB per octave rise with frequency, as shown at Figure 84A. In microphone design, a complementary 6 dB per octave rolloff is incorporated, as shown at Figure 84B, so that the response of the microphone will be flat in a constant pressure field.

In the example shown here, the on-axis dimension of the dipole is such that cancellation takes place at a frequency of about 14 kHz. That dimension is given by:

$$D = 33{,}900/(.5)(14{,}000) = 4.8 \text{ cm}$$

In current design practice for capacitor microphones, a distance about half of this value can be maintained.

References: Boré (1989)
Eargle (1981)

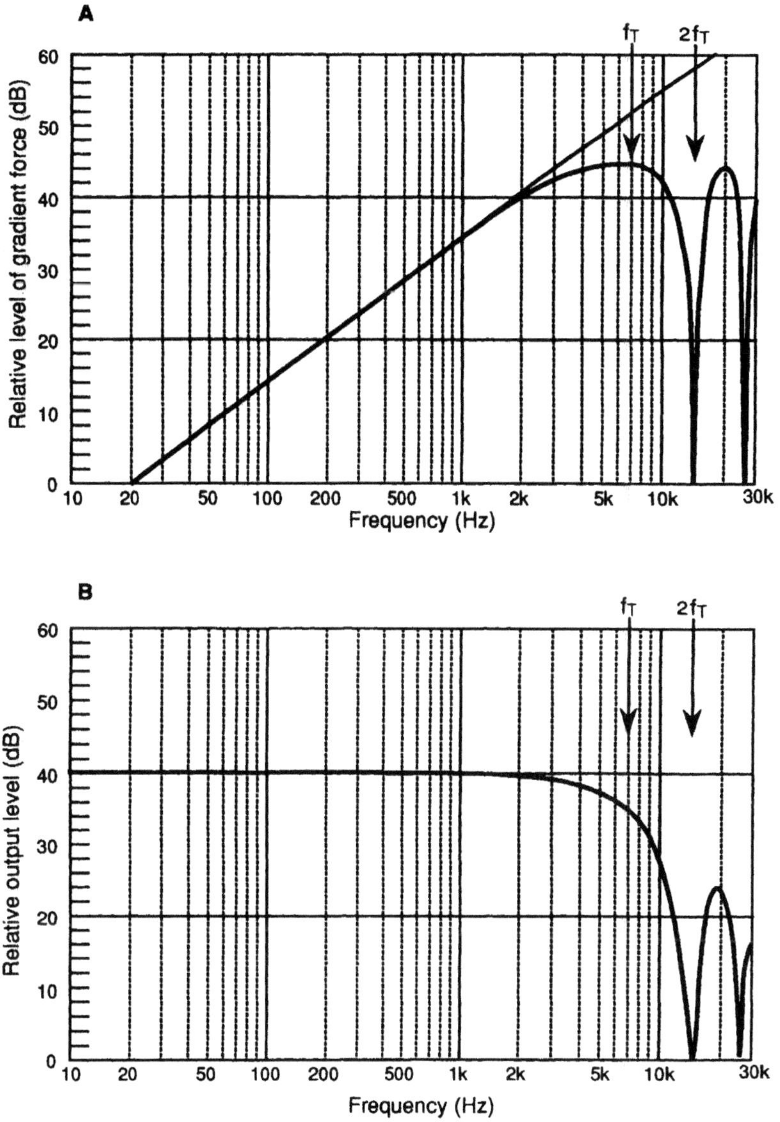

Figure 84. Effect of Dipole Dimension on Directional Microphone Frequency Response

85. Basic Proximity Effect in Directional Microphones

Proximity effect is the tendency of all directional microphones to boost low-frequency response for sources placed close to the microphone. The driving force on the dipole is the difference in sound pressure at the two openings of the dipole; for sources placed at a considerable distance from the dipole, that force will be directly proportional to the driving frequency.

For a sound source located close to the dipole (as shown at Figure 85A and B), there will be a distance differential, ΔD. This will be sufficient to cause an added inverse square pressure difference that will add to the frequency-dependent pressure gradient force.

The inverse square component is constant with frequency, but the frequency-dependent component rises with frequency, as shown in Figure 85C. The net effect is that the inverse square component dominates at low frequencies and causes a rise in the output of the dipole element, as shown in Figure 85D.

References: Boré (1989)
Eargle (1981)

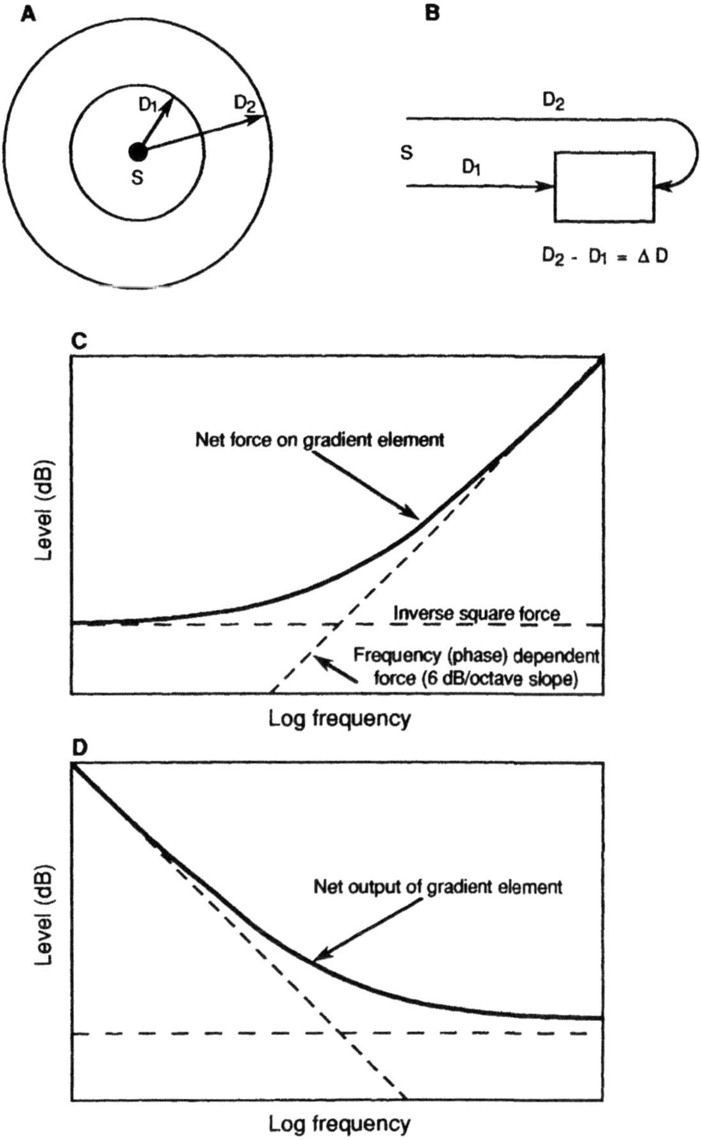

Figure 85. Basic Proximity Effect in Directional Microphones

86. Proximity Effect in a Dipole Microphone at Several Distances

Figure 86 shows the rise in low-frequency response at several working distances for a dipole microphone along its primary axis. The curve is determined by the following equation:

$$\text{Response} = \frac{\sqrt{[1 + (kr)^2]}}{kr} \quad \ldots\ldots\ldots\ldots\ldots\ldots\ldots\ldots (86.1)$$

where: r = distance from the source
k = $2\pi f/c$
c = velocity of sound
f = frequency

The frequency at which inverse square and frequency dependent effects are equal is given by the following equation:

$$\text{Frequency} = c/(2\pi r) \quad \ldots\ldots\ldots\ldots\ldots\ldots\ldots\ldots (86.2)$$

At this frequency, the response will be 3 dB higher than at large distances.

EXAMPLE: For a dipole microphone at an operating distance of 0.5 meter, determine the pressure rise due to proximity effect at 50 Hz.

Locate 50 along the abscissa and read upward to the curve labeled 54 cm, which is approximately 0.5 meter. Then read on the ordinate the value of approximately 7 dB.

References: Boré (1989)
Eargle (1981)

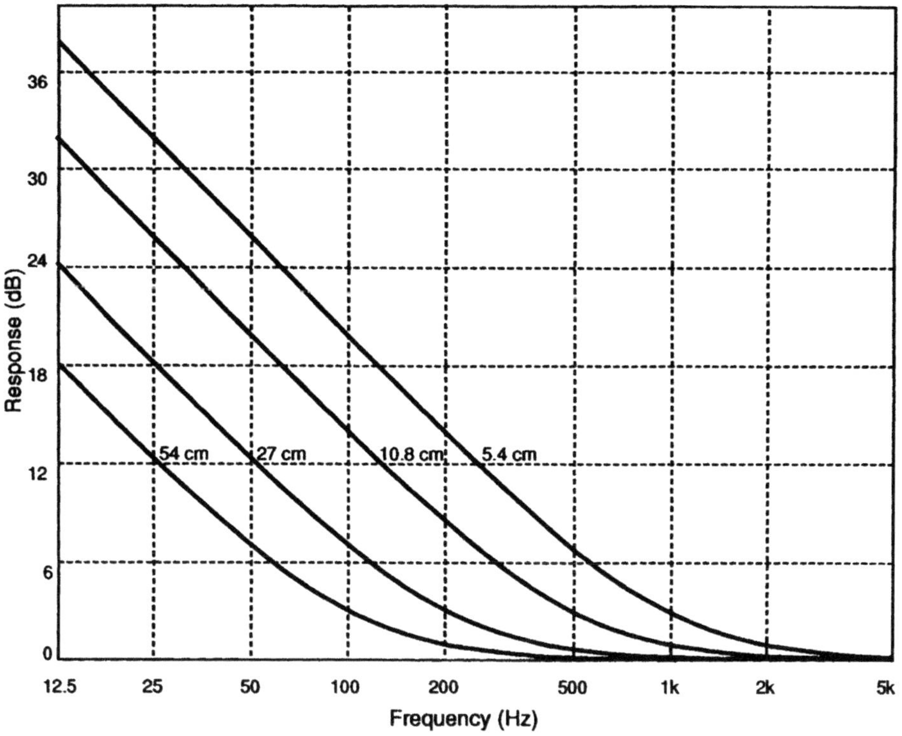

Figure 86. Proximity Effect in a Dipole Microphone at Several Distances

87. On-Axis Proximity Effect in a Cardioid Microphone at Several Distances

For a given working distance and frequency, the on-axis proximity effect in a cardioid microphone will be one-half that of a dipole microphone. The reason is simply that the response of the cardioid microphone is composed of equal amounts of dipole and omnidirectional components, and only the dipole component contributes to the proximity effect.

Thus, the curves shown in Figure 87 are simply those of Figure 86 transposed down one octave.

We can make the following observations for the other patterns in the cardioid family: The supercardioid and hypercardioid patterns will exhibit a *greater* proximity effect than does the normal cardioid, since those patterns consist of greater dipole component. The subcardioid pattern will exhibit a *lesser* proximity effect than the normal cardioid, since it consists of less dipole component.

References: Boré (1989)
Eargle (1981)

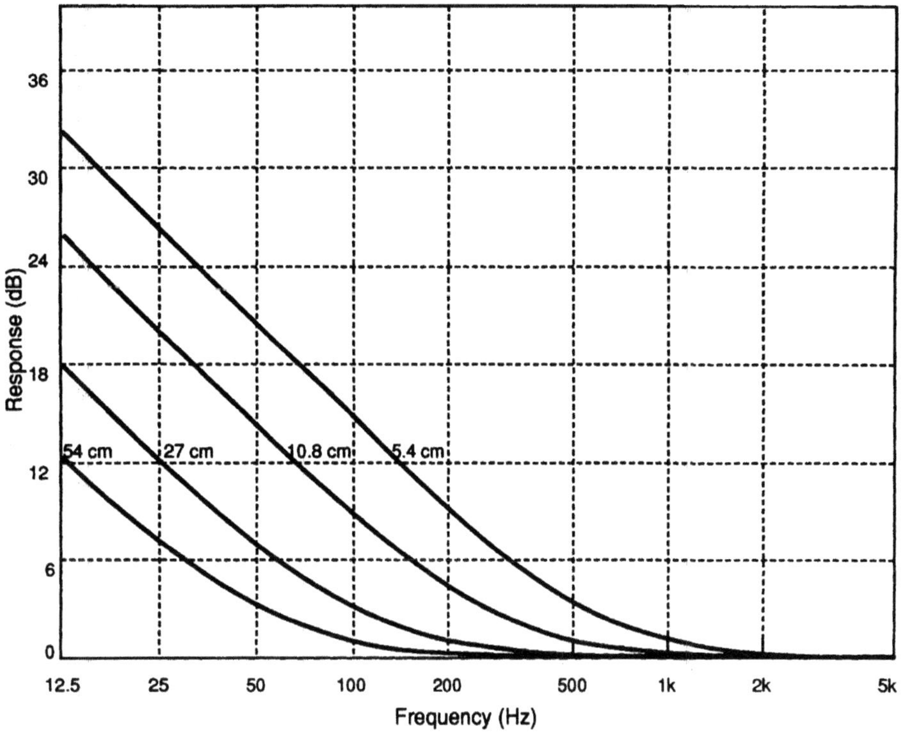

Figure 87. On-axis Proximity Effect in a Cardioid Microphone at Several Distances

88. Proximity Effect in a Cardioid Microphone as a Function of Azimuth Angle

Because the cardioid pattern consists of both omnidirectional and dipole components, it will exhibit varying amounts of proximity effect depending on its angular orientation with respect to the sound source. For example, at 90° off-axis, the dipole component is 0 (see section 79) and there will be no proximity effect.

At an angle of 180°, the response of the cardioid for distant sources is zero, since inverse square and frequency-dependent components are equal and opposite. But for close sources, the frequency-dependent component will rise at low frequencies, producing a very "boomy" effect.

These directional aspects are shown in Figure 88 for a normal cardioid microphone at a working distance of about 0.5 meter.

It is a worthwhile experiment to set up a cardioid microphone in a studio and listen to its response to the spoken voice at several working distances and angles. These experiments will generate a new awareness of what nominal microphone patterns can and cannot accomplish.

References: Eargle (1981)
Olson (1957)

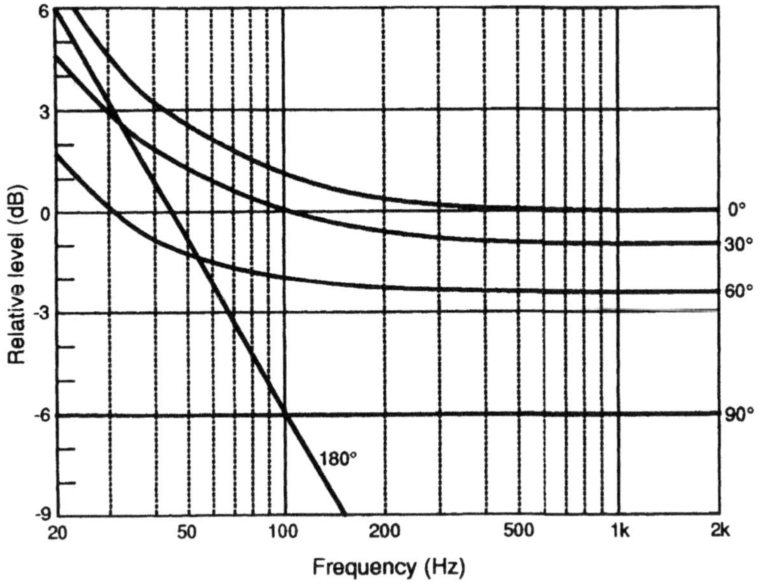

Figure 88. Proximity Effect in a Cardioid Microphone as a Function of Azimuth Angle

89. ON-AXIS AND DIFFUSE FIELD INCIDENCE RESPONSE OF OMNIDIRECTIONAL MICROPHONES

A piston mounted at the end of a long tube exhibits a narrowing of on-axis response at high frequencies (see section 33). Whether as loudspeaker or as microphone, the effect will be the same, taking into account the appropriate scaling due to wavelength and diameter.

With microphones, it is useful to compare the on-axis response in a free field to the response in a diffuse, or random, field, where there is equal likelihood that sound may arrive from any direction. Figure 89 shows the relative response of three microphones in both sound fields.

Microphones of nominal diameters of 25.4 mm, 12.7 mm, and 6.3 mm are represented. The designs shown here have been optimized for flattest on-axis response; their random incidence response shows a characteristic rolloff. Note that for successive halving of the microphone diameter, the frequency response moves upward one octave.

Diffuse field response is not easy to measure, inasmuch as it is difficult to generate a truly diffuse field over a wide band of frequencies. For the most part, a microphone's random incidence response can be approximated by placing that microphone at an angle of 90° relative to a sound source in a free field.

Instrumentation microphones are available with either flat on-axis or flat random incidence response; the user must be sure which is required for a given kind of measurement.

Studio microphones are also available in both kinds. The choice is based on application and microphone diameter. Generally, with small-diameter microphones (12 mm), there will be little practical difference between those that are equalized for flat on-axis and those equalized for flat random incidence. However, with larger-diameter microphones, this difference is transposed downward in frequency and may be more noticeable.

References: Brüel and Kjær (1971)
Eargle (1981)

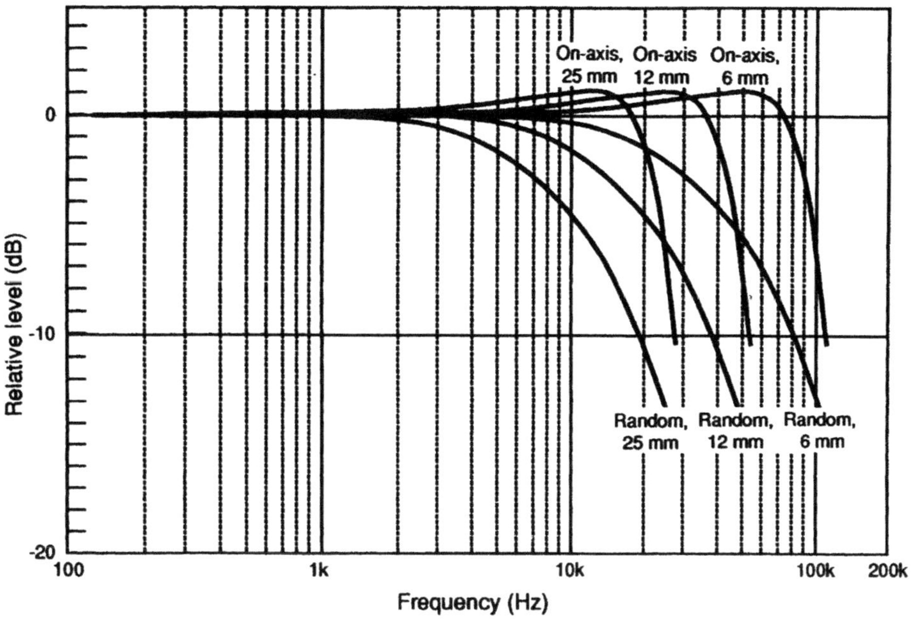

Figure 89. On-axis and Diffuse Field Incidence Response of Omnidirectional Microphones

90. Delay versus Level for Accent Microphones in Recording

The information given here is specific for recording applications where there is a main microphone array for overall stereophonic pickup and one or more accent microphones to accent or highlight a given instrument or section of instruments. Many recording engineers routinely delay the signals from accent microphones in order to compensate for the acoustical delay path between each accent microphone and the main microphone array. When this is done, the contributions of accent and main microphones will be more or less coherent.

As shown at Figure 90A, the most useful range for accent microphones is between –6 and –12 dB relative to the main pickup level. If accent microphones are mixed in at higher levels, their function is more that of solo microphones.

Even though it is always correct to delay accent microphones, it may not be necessary, depending on audibility thresholds. The data shown at Figure 90B were developed by the author as a general guideline.

For relatively high levels of the accent microphone signal, delay is desirable when the accent signal anticipates the main signal by as little as 10 msec. For lower levels of the accent microphone signal, delay may not be necessary until the accent microphone signal anticipates the main signal by as much as about 27 msec. At that point and beyond, the undelayed signal may be perceived as a pre-echo; accordingly, delay compensation is recommended regardless of level.

Reference: Eargle (1992)

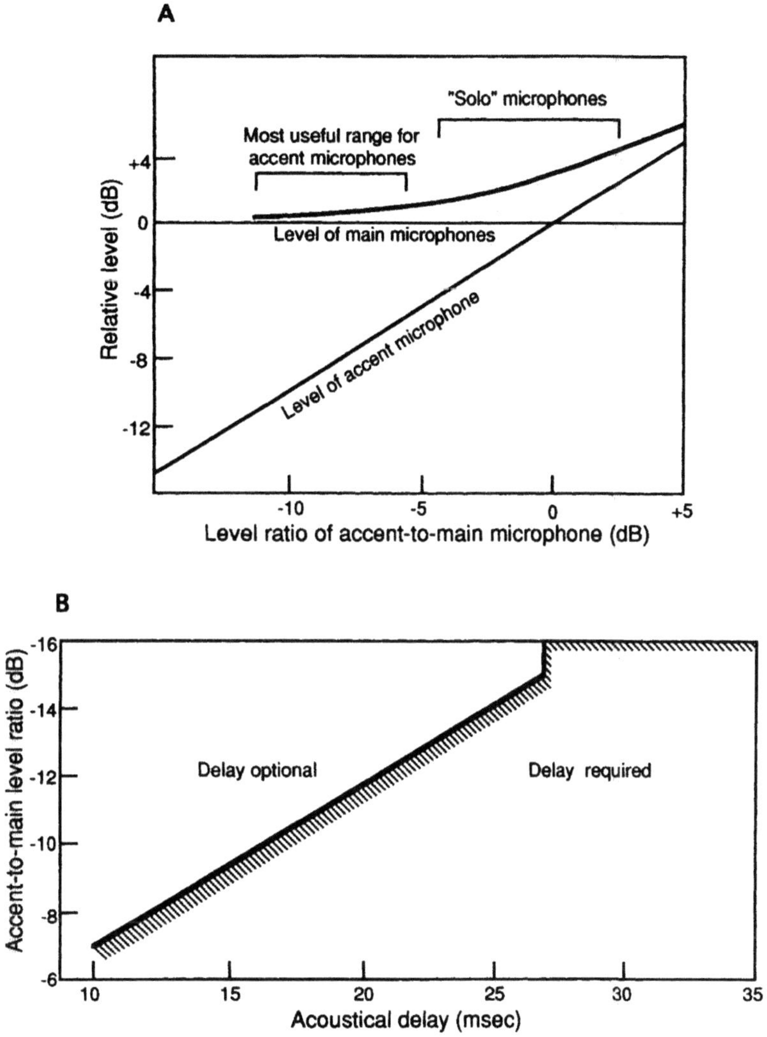

Figure 90. Delay versus Level for Accent Microphones in Recording

91. MICROPHONE BOUNDARY SIZE VERSUS 2π TO 4π TRANSITION FREQUENCY

When boundary layer microphones are operated on the floor or on a wall, the effective acoustical boundary is half-space, or 2π. In many applications, these microphones are, out of necessity, placed on relatively small plane surfaces; at low frequencies, the effective acoustical boundary is full-space, or 4π.

Although the acoustical problem is analogous to that discussed in section 56, the data presented here will, for a fixed boundary perimeter, give an effective transition frequency that is somewhat higher than that indicated by Figure 56. The intent here is to define the smallest boundary perimeter that will ensure smooth frequency response.

Reference: Eargle (1992)

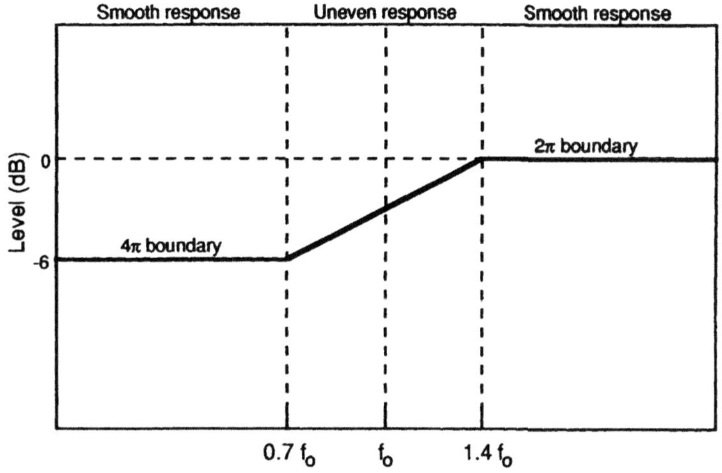

Boundary diameter:	f_o:
15 meters	30 Hz
9 meters	50 Hz
4.5 meters	100 Hz
1.5 meters	300 Hz
0.9 meters	500 Hz
0.45 meters	1000 Hz

Figure 91. Microphone Boundary Size versus 2π to 4π Transition Frequency

92. Higher-Order Microphone Characteristics

The basic directional microphone patterns are referred to as "first-order patterns," inasmuch as the directional response is proportional to the first power of a cosine term. A second-order pattern has directivity that is proportional to the square of the cosine term. Stated differently, a first-order microphone has response that is proportional to the pressure gradient, whereas a second-order design has response that is proportional to the gradient of the gradient.

First-order gradient microphones require that the sound field be sampled at two points. A second-order gradient microphone is normally composed of two first-order gradient microphones; thus, they will occupy about twice the space of a single first-order microphone. As a result higher-order microphones are usually frequency-limited in their gradient-dependent response.

The primary use for higher-order gradient microphones is in speech applications, where greater immunity to noise is always an advantage. Such microphones may be simulated in the field by combining first-order gradient microphones at small distances and connecting one of them in reverse polarity. In this case, considerable bass boost will be required for flat response of distant sound sources.

Figure 92A shows the response of a single-order bidirectional microphone. Polar equation, distance factor, and random energy efficiency (REE) are given. Second- and third-order bidirectional patterns are shown at Figure 92B and C. Note the considerable increase in distance factor as the order increases. Figures 92D and E show similar data for second-order unidirectional patterns that offer even greater distance factor.

A study of the references will underscore the problems inherent in these designs.

References: Beavers and Brown (1970)
Olson (1957)

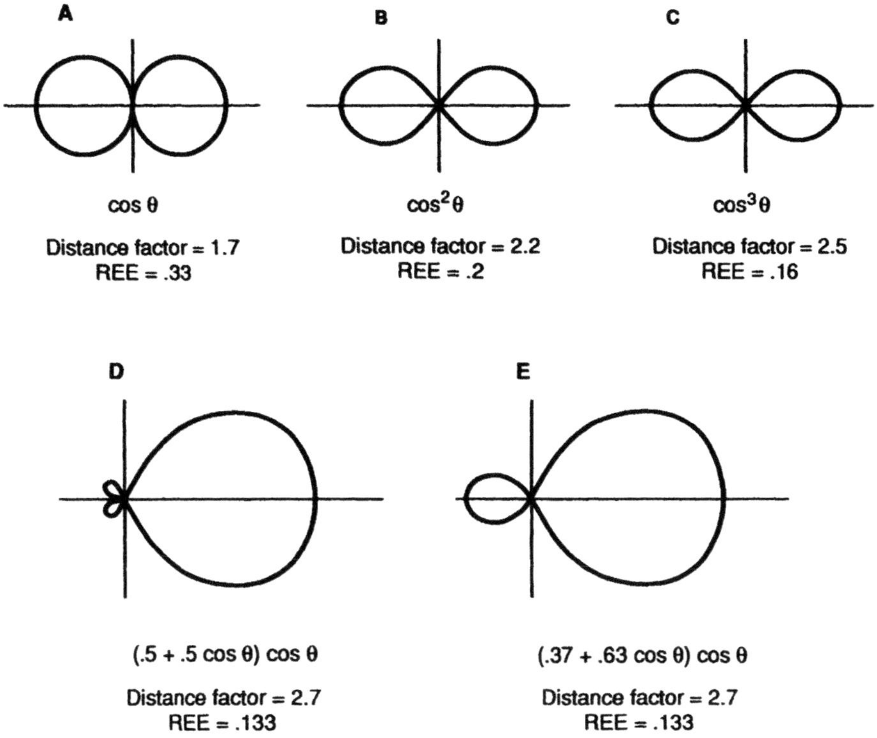

Figure 92. Higher-Order Microphone Characteristics

93. Microphone Line Losses

Figure 93 shows the effects of high-frequency loss in microphone cable with a capacitance between the conductors of 100 picofarads per meter of cable length and resistance of 0.08 ohms per meter. These values are typical for microphone cable.

It is clear that microphones with a source impedance of 200 ohms or less will exhibit negligible high-frequency losses for line runs up to 60 meters. For longer runs, microphone preamplifiers should be placed at the microphone and that line level should be implemented using amplifiers with sufficiently low source impedance.

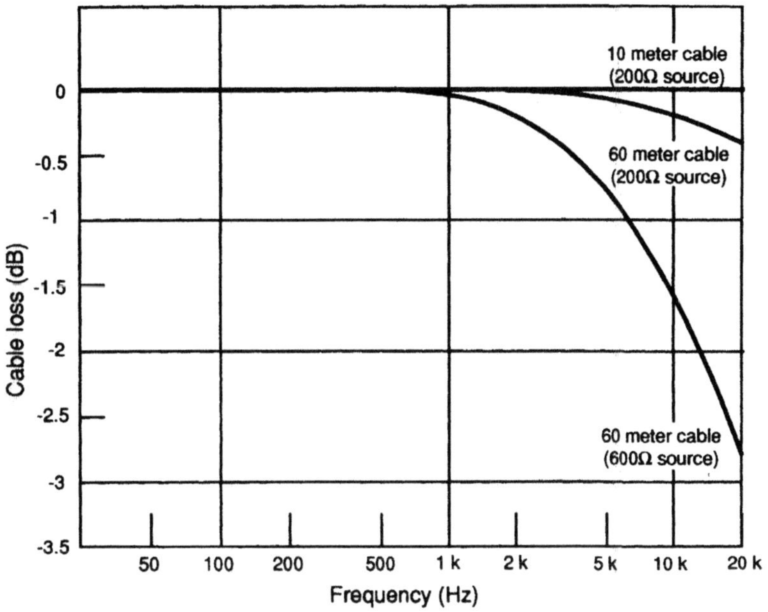

Figure 93. Microphone Line Losses

PART FOUR

SIGNAL TRANSMISSION

94. Time Constant versus Frequency

In many applications dealing with signal pre-emphasis and de-emphasis, engineers speak of the time constant of a resistor-capacitor combination instead of transition frequency. The relation between the two is:

$$\tau = 1/2\pi f \qquad (94.1)$$

$$f = 1/2\pi\tau \qquad (94.2)$$

where τ is the time constant in seconds.

The usefulness of the concept is that in designing equalization networks, the engineer can simply use a resistor-capacitor combination of the requisite value of time constant. The time constant of a resistor-capacitor combination is:

$$\tau = RC \qquad (94.3)$$

where R and C are, respectively, the values of resistance in ohms and the capacitance in farads.

Frequencies from 10 to 1000 Hz are related to time constant along the left ordinate of Figure 94, whereas those from 1000 to 100 kHz are along its right ordinate.

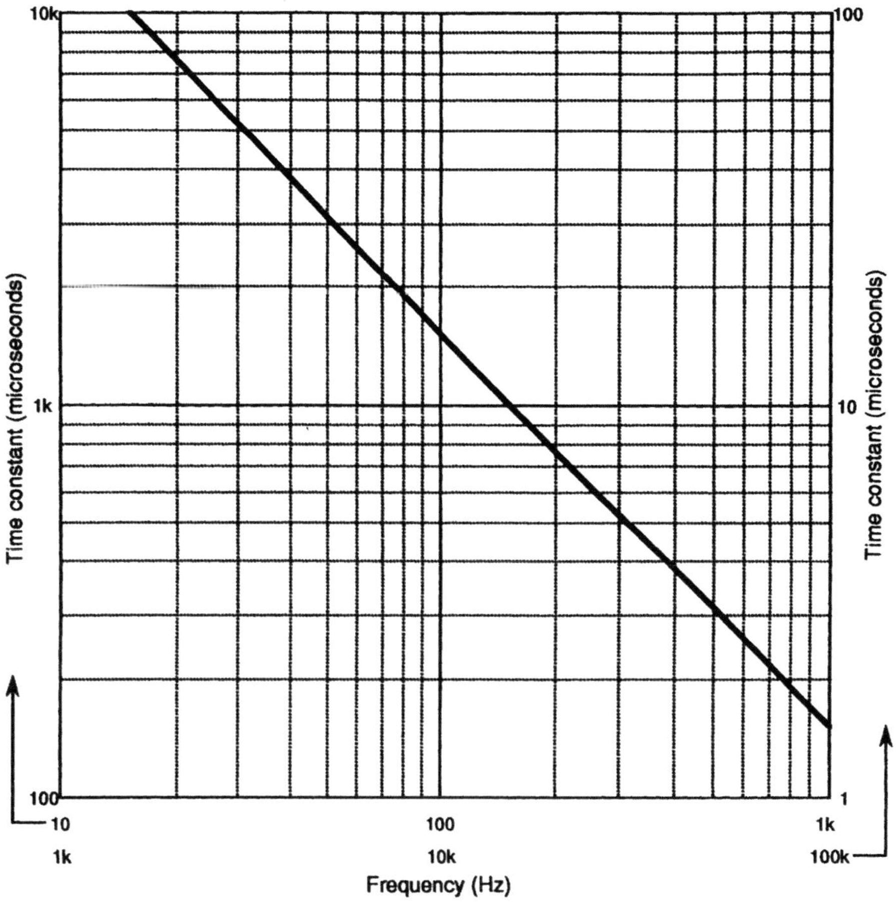

Figure 94. Time Constant versus Frequency

95. RIAA (Record Industry Association of America) Disc Pre-emphasis and De-emphasis

Both mono and stereo 33 $\frac{1}{3}$-rpm discs are recorded and played back with electrical pre- and de-emphasis established by the RIAA (Record Industry Association of America) during the 1950s. These equalization curves were subsequently adopted internationally, to the exclusion of all other equalization schemes.

The RIAA curves have time constants at 3180, 318, and 75 microseconds, which correspond to transition frequencies of 50, 500, and 2120 Hz. In Figure 95, A represents the electrical equalization for a signal that is to be fed to a constant velocity cutting system, whereas B represents the electrical equalization of a signal received from a velocity responsive pickup, such as a moving coil or moving magnet type.

The overall high-frequency boost in recording is necessary to keep the recorded signal well above the noise level of the medium; the signal reduction at lower frequencies is necessary to preserve valuable modulation space on the disc. The high-frequency boost above 2120 Hz requires careful monitoring of high-frequency content in the program material; in many instances, high-frequency signal limiting is required if overall recorded level is to be maintained.

As a practical matter, the RIAA recording specification is normally taken out to 20 kHz, with additional low-pass filtering above that frequency to prevent system overload. On the playback side, a rolloff is often introduced below 20 Hz in order to diminish the effects of record warp, low-frequency granularity noise, and electromechanical feedback.

Reference: Eargle (1992)

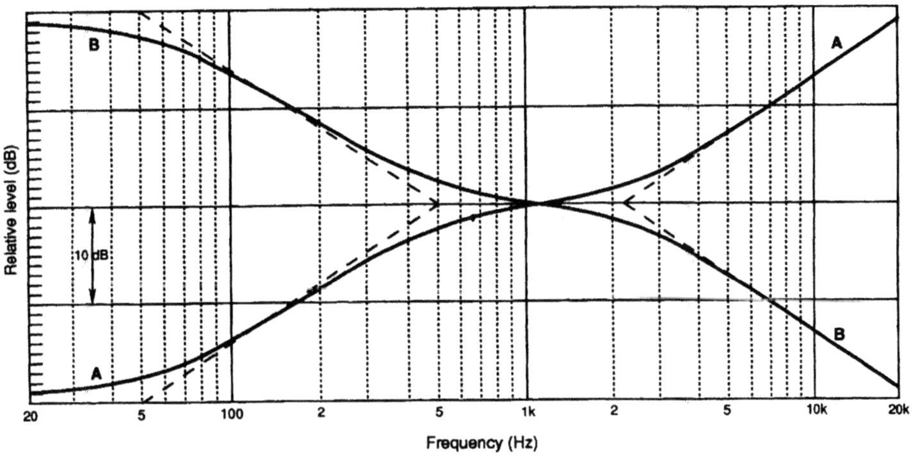

Figure 95. RIAA Disc Pre-emphasis and De-emphasis

96. FM Broadcasting Pre-emphasis and De-emphasis

With the coming of FM stereo in the 1950s, the added transmission requirement of a subcarrier L − R channel brought with it a considerable increase of high-frequency noise. In order to compensate, the standardized system included a high-frequency pre-emphasis with a 75-microsecond time constant at the broadcast transmitter and a complementary de-emphasis at the receiving end.

In Figure 96, the pre-emphasis is shown at A and the de-emphasis at B. The standard is defined out to 15 kHz.

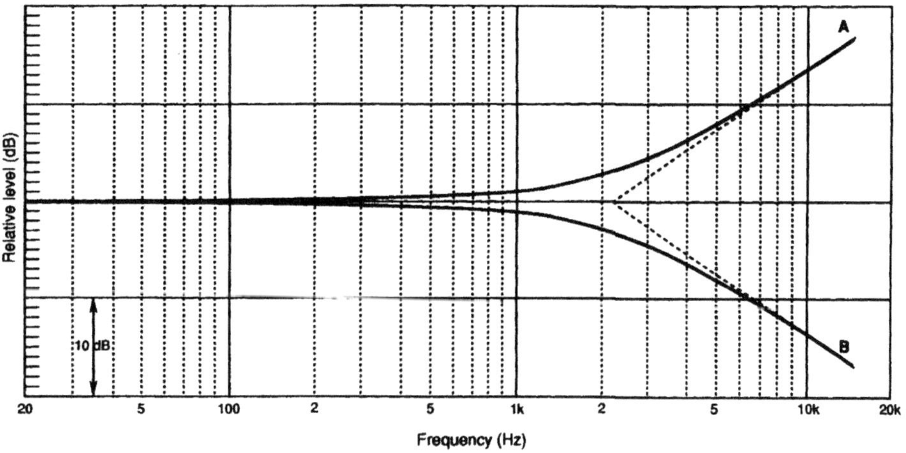

Figure 96. FM Broadcasting Pre-emphasis and De-emphasis

97. Early 78-rpm and $33\frac{1}{3}$-rpm Disc Pre-emphasis and De-emphasis Standards

Figure 97 shows the frequency breakpoints for the multiplicity of disc playback standards that existed during the 1940s and early 1950s. It is clear that there was little difference between many of these breakpoints. A single standard could be defined that would not be more than, say, 2 dB removed from most of them. This was the starting point for the definition of the RIAA standard.

Basically, the RIAA standard adopted a high-frequency rolloff that was about halfway between the values for the AES and the NAB-Columbia standards. The NAB-Columbia low-frequency boost commencing at 500 HZ was adopted, and a final breakpoint at 50 Hz was adopted as a compromise among the Columbia, NAB, and AES standards.

Reference: Newitt (1953)

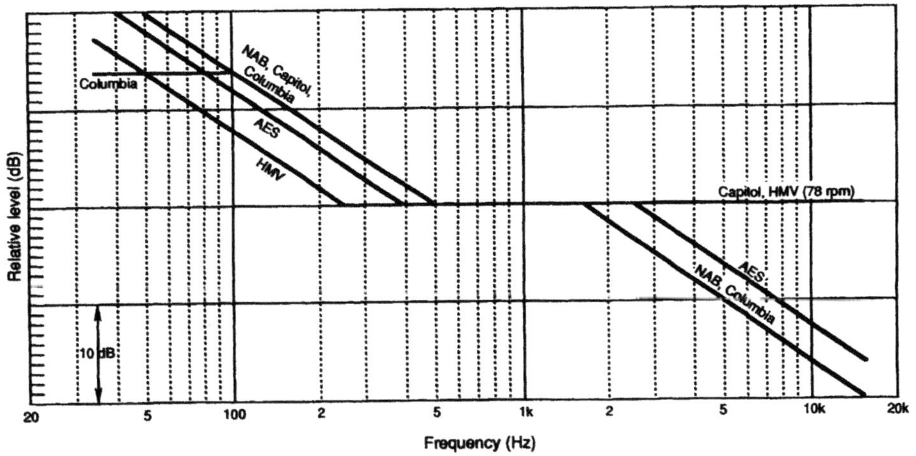

Figure 97. Early 78-rpm and 33⅓-rpm Disc Pre-emphasis and De-emphasis Standards

98. Motion Picture Mono Optical Reproduce Standard

As optical recording developed for 35-mm motion pictures, both record and playback were conceived as flat processes. On the record side, it was necessary to correct high-frequency resonances in the galvanometer light modulator; on the playback side, it was necessary to add some high-frequency correction for slit, or scanning, losses. Each of these was highly dependent on the particular hardware that was used.

Because of noise due to film emulsion granularity, optical prints were not generally played back with a flat reproduce characteristic. In fact, the Theater Standardization Committee of the Research Council of the Academy of Motion Picture Arts and Sciences established the curve as shown in Figure 98 for rolling off high-frequency noise artifacts that accumulated during all recording and processing steps. The curve is often referred to as the "Academy rolloff," and it is still used in theaters when older monophonic optical prints are shown.

The playback standard was ultimately retired from modern operations when Dolby Laboratories developed the stereo optical format, where Dolby A-type noise reduction was used to remove noise through complementary code-decode processes. Today, through careful attention to alignment and control of slit width, it is possible to achieve optical performance beyond 12.5 kHz in the theater.

Reference: Frayne and Wolfe (1949)

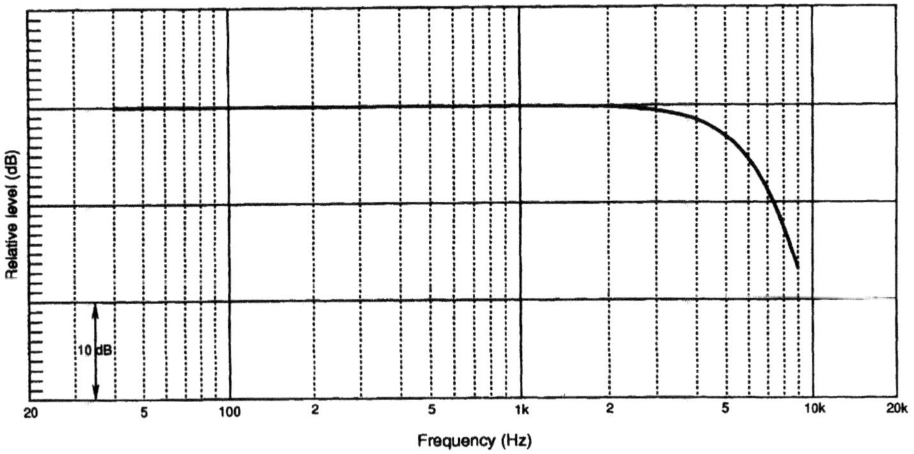

Figure 98. Motion Picture Mono Optical Reproduce Standard

99. Digital Pre-emphasis and De-emphasis Standard

In the early development of digital standards for both professional and consumer applications, 14-bit systems were considered. In order to make the 14-bit standard workable for low-level music, a pre-emphasis and complementary de-emphasis scheme was implemented with time constants of 50 and 15 microseconds. These values correspond to transition frequencies of 3180 Hz and 10,600 Hz, as shown in Figure 99. A recording that has been made with pre-emphasis carries a flag that will tell the playback system to switch in the complementary high-frequency rolloff.

For speech and most musical requirements, the high-frequency boost in recording can be accommodated without any essential loss in overall system power bandwidth. However, certain kinds of music with accented high-frequency content can pose problems.

In today's 16-bit digital world, the standard is not really necessary; however, most recording and playback systems still provide it as a user option.

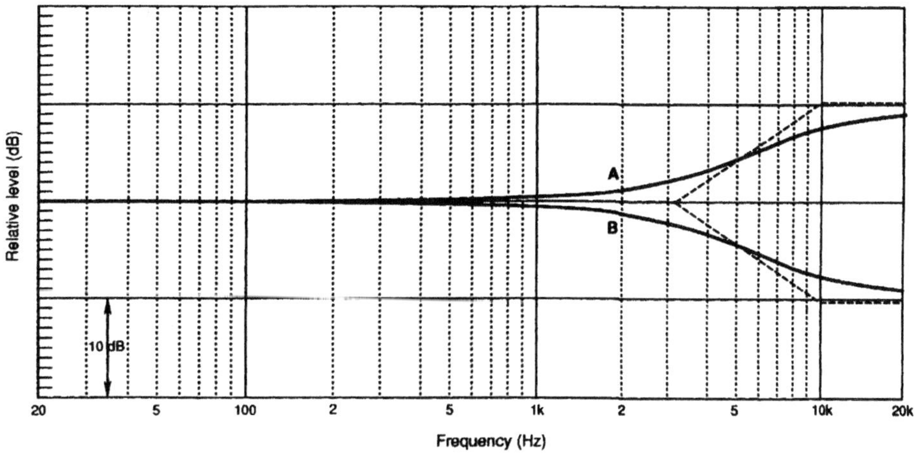

Figure 99. Digital Pre-emphasis and De-emphasis Standard

100. Comparison of Meters Used in Broadcasting and Recording

Figure 100A shows comparative scales of three kinds of IEC (International Electrotechnical Commission) PPM (peak program meters) and the VU (volume unit) meter commonly used in the United States. These meters normally have scales that are curved, but they are shown in linear form here to facilitate comparisons between them.

The various PPM are electronically driven and have attack and release times that can be tailored as desired. The VU meter is traditionally a passive device; as such, it has fairly slow attack and release times that are fixed.

The scales of the meters are lined up so that the normal signal alignment or reference levels appear one over the other. For most activities involving analog media, the normal maximum signal level is generally taken to be 9 dB above the reference level, as indicated in Figure 100A.

As a group, the PPM have similar ranges, but they exhibit different scale numbering. They also show a greater range of signal activity above the nominal alignment level than does the VU meter. In general, the alignment point is about at the middle of the scale, with +12 dB and −12 dB around it. The BBC (Type IIa) meter shows this in the most elegant way, with simple markers numbered 1 through 7 at 4-dB intervals.

The attack characteristics of the various meters are shown at Figure 100B and the release times at Figure 100C. Both attack and release time constants of the VU meter are the same. For the PPM, the attack times are such that the meters will reach their full indication in about 10 milliseconds; their decay times are of the order of 2–3 seconds so that the peak values can be more easily observed.

Because of these ballistic differences, the VU meter tends to register lower levels on the same program material than the various PPM. For speech, this difference, sometimes called "lead factor," is about 6–8 dB. For a given recording or transmission medium, it is customary to modulate levels by a corresponding amount. In magnetic recording, for example, engineers using VU meters will use a reference tape fluxivity of 250 nanowebers per meter; whereas the PPM user, with the same tape stock, will use a reference tape fluxivity of 510 nanowebers per meter, which is about 6 dB higher. When this is done, similar meter deflection will result in similar actual recorded levels on the tape.

References: Eargle (1992)
Thiele (1985)

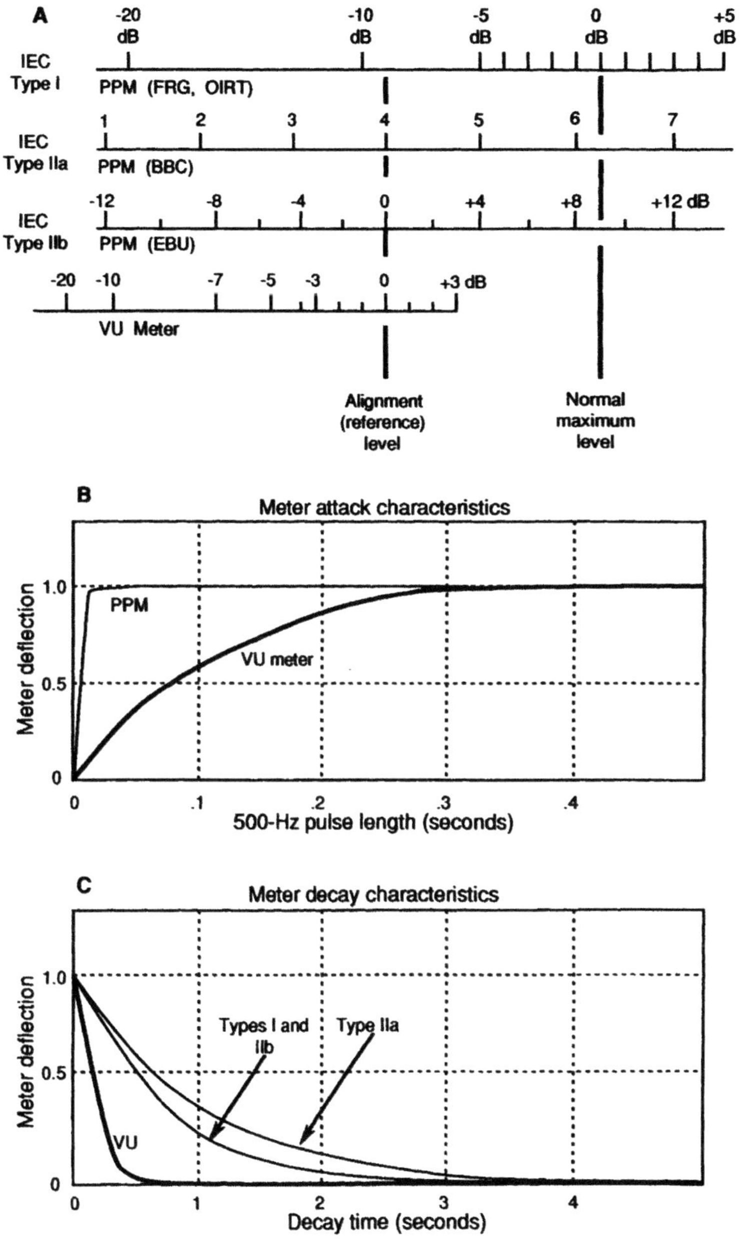

Figure 100. Comparison of Meters Used in Broadcasting and Recording

101. Power Ratios Expressed in dBm

In line level signal transmission, the term "dBm" (dB re 1 milliwatt) is often encountered. Fundamentally, a power level in dBm is defined as:

Level (dBm) = 10 log (P/.001) 101.1

where P is the power in watts.

In telephony and broadcasting, which were the precursors of modern audio transmission, so-called 600-ohm lines were used. What this indicates is that all devices in the transmission chain had both input and output impedances of 600 ohms; thus, the signal level could be conveniently expressed in terms of milliwatts and in levels relative to 1 milliwatt.

The scale shown in Figure 101 extends from −30 to +30 dBm, corresponding to powers of .001 milliwatt to 1 watt. However, we do not normally measure power directly. We measure voltage, while calibrating the meter scale in watts or milliwatts as appropriate.

For an actual 600-ohm transmission system, the voltage corresponding to 1 milliwatt is given by the following equation:

$$E = (600 \times .001)^{1/2} = 0.775 \text{ volts}$$

EXAMPLE: An audio mixing console has a published maximum power output capability of 20 dBm. What is the maximum possible output voltage when the console is loaded with 600 ohms?

We must watch the wording here carefully. The specification implies that the console can be loaded with 600 ohms, because this is the traditional load associated with the dBm scale. And since a voltage of 0.775 volts corresponds to 0 dBm, a level 20 dB greater will correspond to a voltage of 7.75 volts. (Remember that a 10-to-1 voltage ratio corresponds to a level difference of 20 dB.)

Section 102 will make the picture clearer.

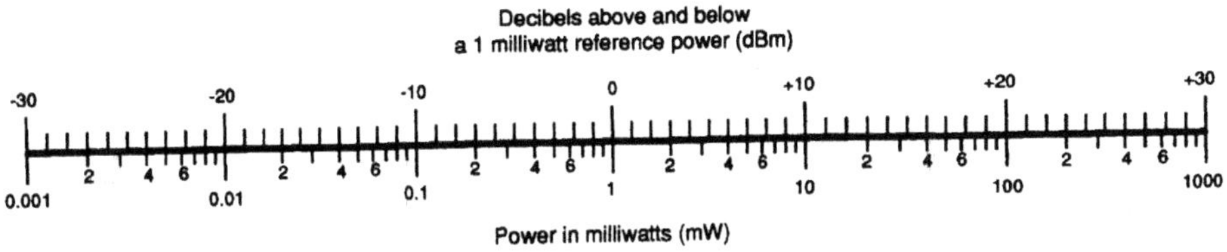

Figure 101. Power Ratios Expressed in dBm

102. VOLTAGE RATIOS EXPRESSED IN dBu

The term "dBu" refers to voltage levels relative to 0.775 volts, regardless of load impedance. It is basically a holdover from an earlier era when power levels in signal transmission systems were expressed in dBm. The fundamental definition is:

Level (dBu) = 20 log (E/.775) .... (102.1)

where E is the voltage in volts.

As voltage transmission systems became common, the older measurement standard was maintained. Today we commonly see signal levels in all branches of audio technology and engineering expressed in dBu. The designation "dBv" is sometimes used, but it is likely to be confused with dBV, which is based on another reference.

EXAMPLE: A recording console has an output capability of 22 dBu. What will be the maximum possible output voltage?

Consult Figure 102 and locate +22 on the dBu scale. Read approximately 10 volts on the opposite side.

Note: When consoles and other items of audio electronics are rated only in dBu, it is assumed that they are *not* to be loaded with 600 ohms. In general, these devices are to be operated effectively open circuit, or into a load that is 3000 ohms or higher in impedance.

EXAMPLE: A microphone has an internal impedance of 200 ohms, and its open-circuit output in a 94 dB L_p sound field is specified as −40 dBu. (a) What is its open-circuit output voltage? (b) What will be the output voltage level if the microphone is loaded with 200 ohms?

(a) Find −40 on the nomograph and read 8 millivolts on the opposite side. (b) When a voltage source is loaded with an impedance equal to its internal impedance, the voltage will drop by half. By inspection, we can see that the level will be 6 dB lower, or −46 dBu.

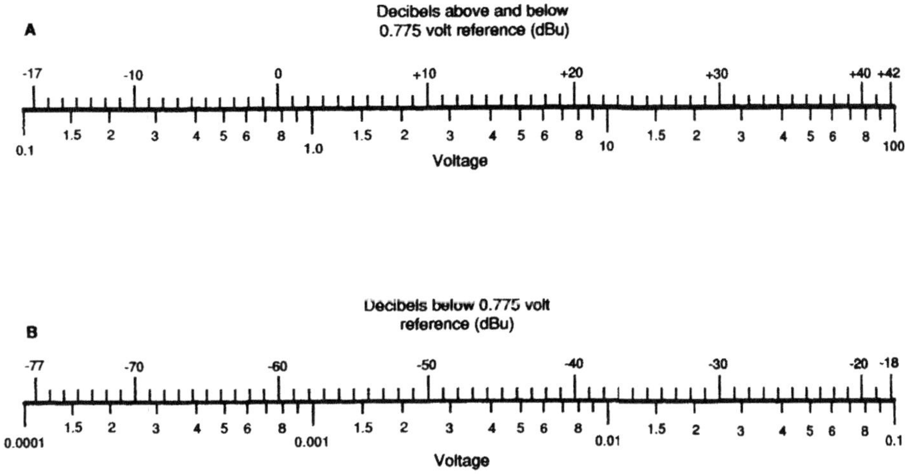

Figure 102. Voltage Ratios Expressed in dBu

103. Power Ratios Expressed in dBW

The reference for this measurement is 1 watt; the definition is:

Level (dBW) = 10 log (P) . . . . (103.1)

where P is power in watts.

Levels in dBW are often used in power calculations in the range above 1 watt and in amplifier specifications.

EXAMPLE: It has been determined that a 60-watt amplifier will produce a level of 100 dB L_p with a given loudspeaker at a given distance. A reference level of 108 dB L_p is desired. What size amplifier is needed?

Locate 60 watts on the nomograph in Figure 103 and read the value of about 18 dBW on the opposite side. Add 8 dB to the level to get 26 dBW. Then read the power of 400 watts on the opposite side.

EXAMPLE: The Hoover Dam hydroelectric plant has an output capability of about 1,300,000 kilowatts. Express this as a power level in dBW.

Our nomograph does not cover this range, so we will have to use equation 103.1. Simply enter the output power and solve:

Level (dBW) = 10 log(1.3×10^9) = 91 dBW

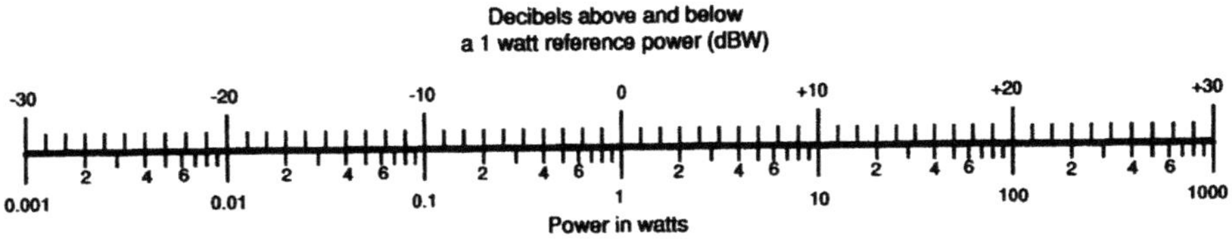

Figure 103. Power Ratios Expressed in dBW

104. Voltage Ratios Expressed in dBV

The reference for this measurement is 1 volt; the definition is:

Level (dBV) = 20 log (E) . . . . (104.1)

where E is the voltage in volts.

EXAMPLE: Find the voltage level in dBV of .001 volts.

On the nomograph in Figure 104B locate the value .001 and read the level of –60 dBV on the opposite side.

EXAMPLE: Compare this result with the voltage level in dBu of .001 volts.

On the nomograph in Figure 102, locate the value .001 and read the level of approximately –57.8 dBu on the opposite side.

These examples show that levels stated in dBu and dBV differ from each other by a relatively small amount.

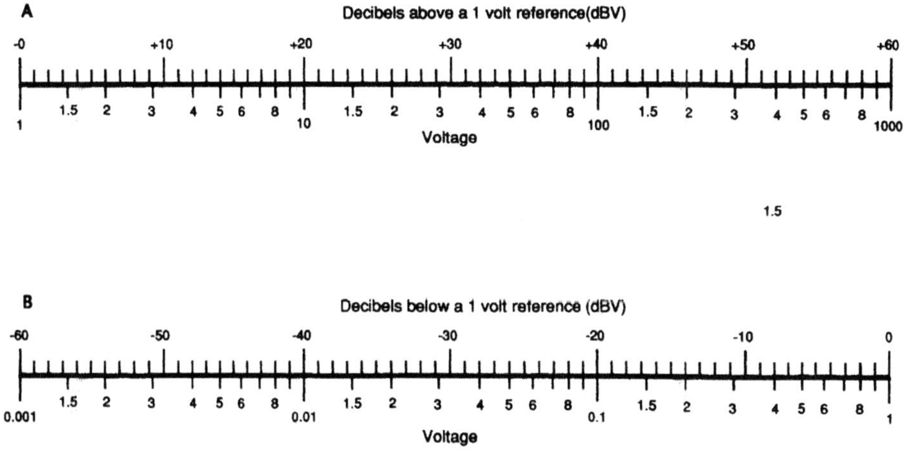

Figure 104. Voltage Ratios Expressed in dBV

105. Sine Wave Voltage Output versus DC Voltage Capability

Figure 105 shows values of dc voltage capability along the abscissa and rms values of voltage for a sine wave signal that can be accommodated by the plus and minus dc voltages. Rms voltage is shown on the left ordinate; average sine wave power into both 4- and 8-ohm loads is shown along the right ordinate.

EXAMPLE: A solid state power amplifier operates with rail voltages of ±40 volts dc. Find the maximum value of average power that the amplifier can deliver to a 4-ohm load.

Locate the dc value of 40 volts along the abscissa. Read upward to the bold line and across to the left ordinate. Read the value of 28 volts rms. Then move across to the 4-ohm ordinate at the right and read the value of 200 watts.

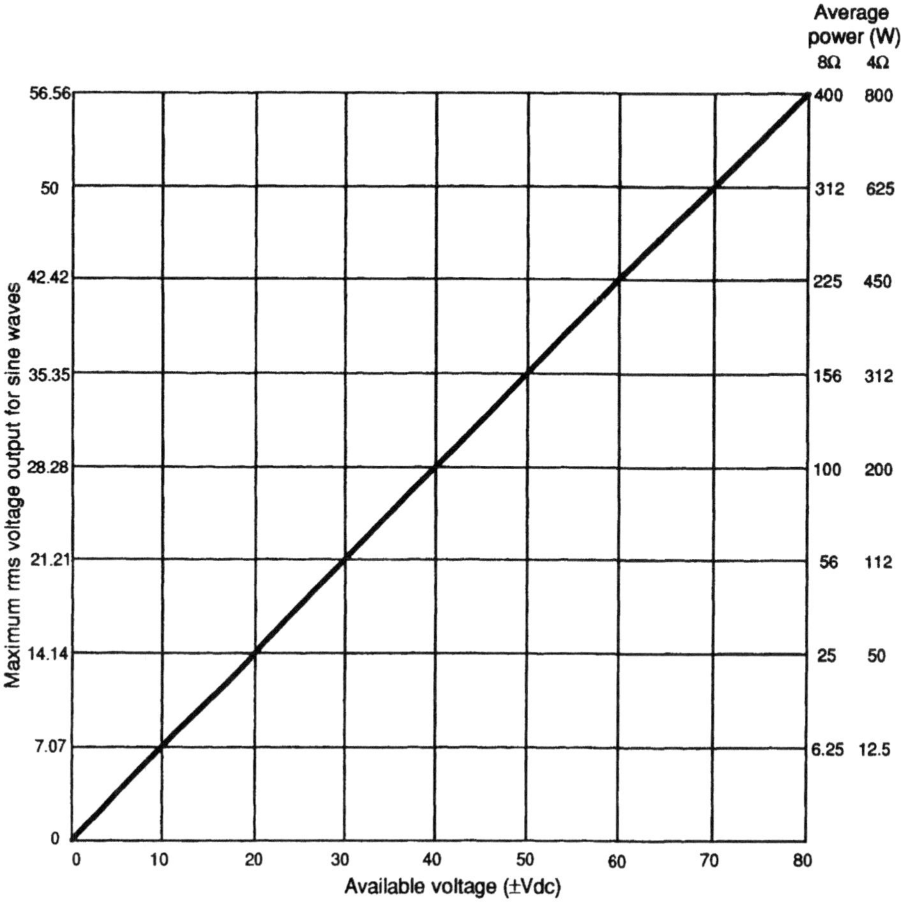

Figure 105. Sine Wave Voltage Output versus DC Voltage Capability

106. Resistance Values for Various Lengths and Gauges of Copper Wire

Wire resistance can be expressed by the following equation:

$$R = kl\,(10^{AWG/10}) \quad (106.1)$$

where l is the wire length in meters and k is a constant equal to .00033 for copper and .00053 for aluminum.

A convenient rule of thumb is that 1000 feet of AWG no. 10 copper wire has a resistance of 1 ohm. For each increase in the gauge by two, the resistance increases by a factor of 1.6.

EXAMPLE: Find the resistance of a piece of AWG no. 30 copper wire 10 meters long.

Along the abscissa in Figure 106, locate the value of 10 meters and move upward to the bold line marked AWG 30. Then move to the ordinate and read the value of about 3.5 ohms. Checking this value with equation 106.1, we get the exact value of 3.3 ohms.

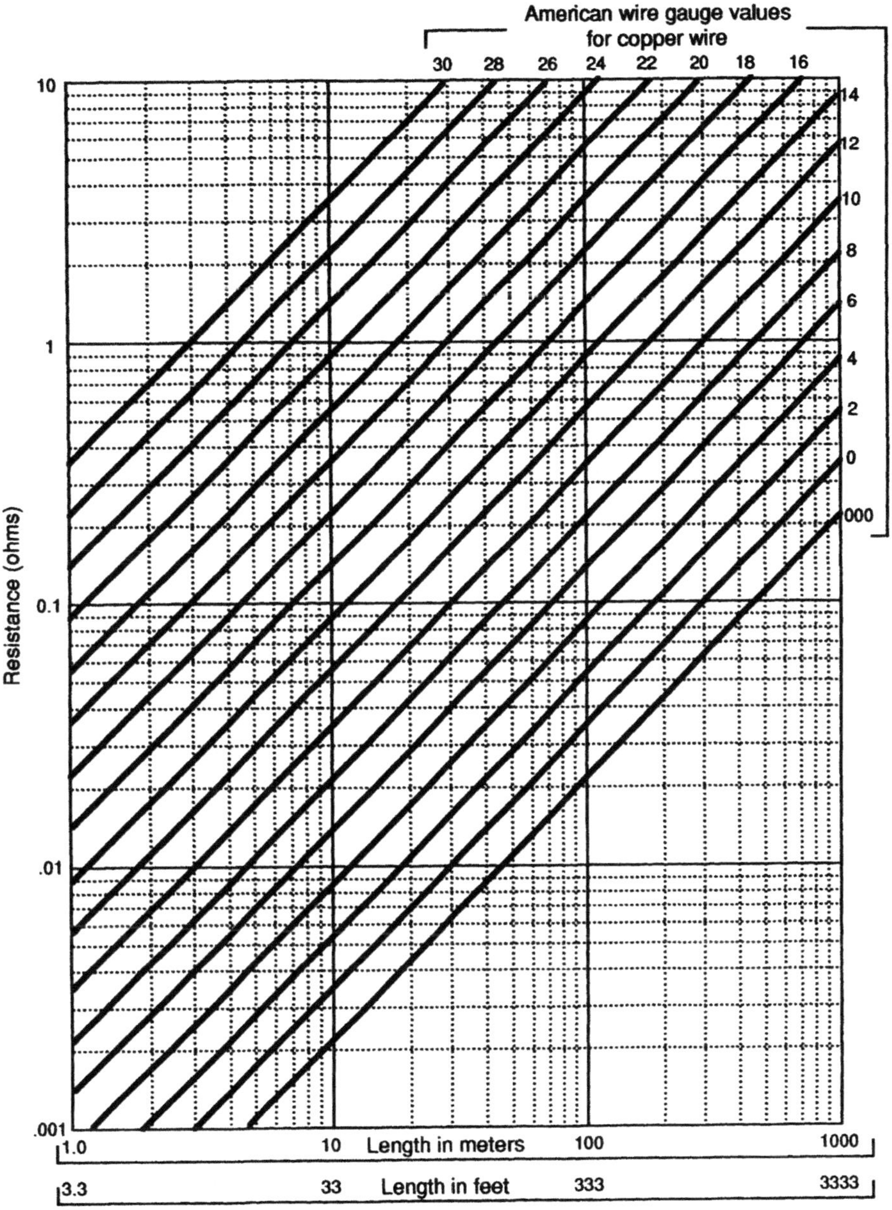

Figure 106. Resistance Values for Various Lengths and Gauges of Copper Wire

107. Metric Wire Gauges

In most metric countries, wire gauge is expressed directly in cross-sectional area in square millimeters. Figure 107 shows the relationship between wire gauge and square millimeters. It is convenient to remember that AWG no. 10 corresponds to 5.2 square millimeters, and each increase in two AWG numbers corresponds to multiplying the area in square millimeters by .63.

It is also convenient to remember that each *increase* of AWG by a factor of 10 corresponds to a *decrease* of one-tenth in cross-sectional area.

When converting metric specifications to AWG and vice versa, all rounding-off errors should be made to the next smaller AWG or the next larger area.

AWG number	mm^2
30	.052
28	.083
26	.13
24	.2
22	.33
20	.52
18	.83
16	1.3
14	2
12	3.3
10	5.2
8	8.3
6	13
4	20
2	33
0	52

Figure 107. Metric Wire Gauges

108. HIGH-FREQUENCY TRANSDUCER PROTECTION CAPACITORS

It is customary in bi- and tri-amplified systems to place a capacitor in series with the amplifier output and the driver in order to protect the driver from inadvertent dc turn-on transients. The capacitance value normally chosen for this has a capacitive reactance equal to the nominal driver impedance at a frequency that is one-half the nominal crossover frequency. This value of capacitance is a good compromise between providing adequate protection and not interfering with the target crossover point.

Because crossover points are often close to the primary resonance of high-frequency drivers, the nominal blocking capacitor value, as given in Figure 108, should be checked by inspecting the published impedance curve for the driver to determine its actual impedance at the intended crossover frequency.

EXAMPLE: A 16-ohm high-frequency driver is crossed over at 500 Hz. Determine the size of a capacitor that will afford proper protection.

The following equation is used for this calculation:

$$C\ (\text{farads}) = \frac{1}{2\pi(16)(250)} = 40\ \mu\text{F}$$

This value may be checked with the graph.

References: Eargle (1989)
JBL (1984)

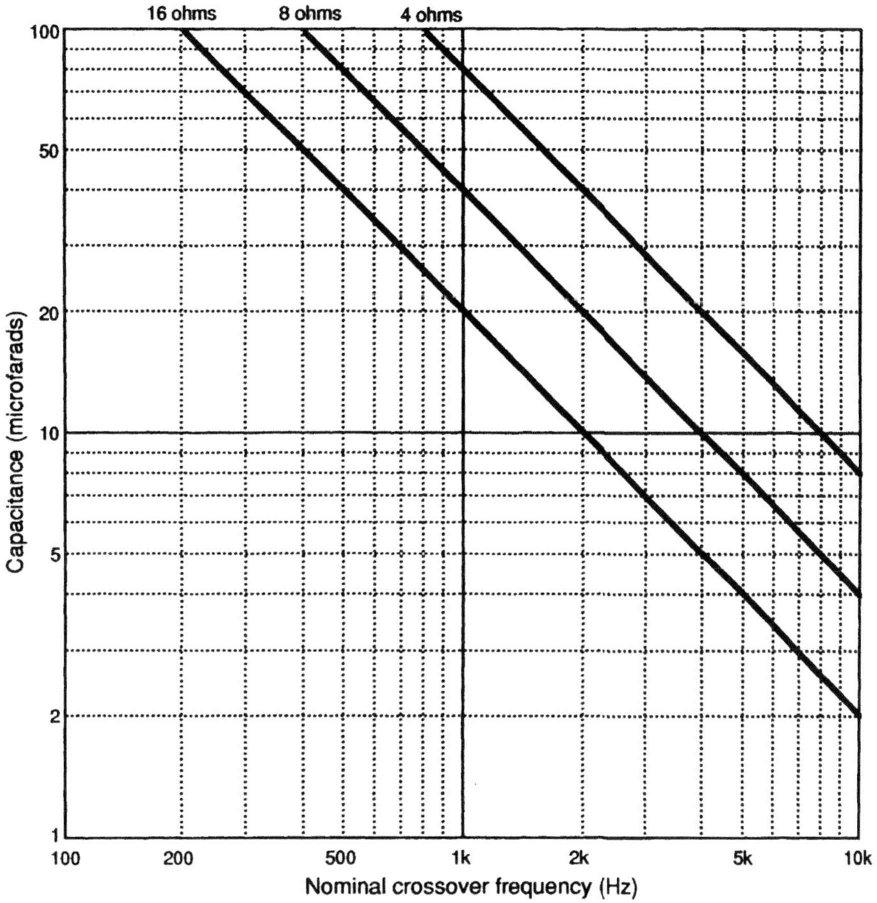

Figure 108. High-Frequency Transducer Protection Capacitors

109. Design of Symmetrical T-Pads

T-pads are useful in so-called matched impedance transmission systems to insert attenuation without upsetting the normal impedance relationships. The symmetrical T-pad can be designed for a given loss while maintaining the design impedance value looking into the pad as well as looking into the adjacent inputs and outputs of other equipment. The arrows in the schematic of Figure 109 make this clear.

In the table shown in Figure 109, the design impedance is 100 ohms. To convert to any other impedance, simply multiply the values of resistance by Z/100, where Z is the design impedance.

EXAMPLE: Determine values R_1 and R_2 for a T-pad that will provide 10 dB loss in a 600-ohm transmission system.

Locate the values of R_1 and R_2 that correspond to a 10-dB loss. Multiply these values by 600/100 = 6 to arrive at values of 312 and 422.4 ohms, respectively.

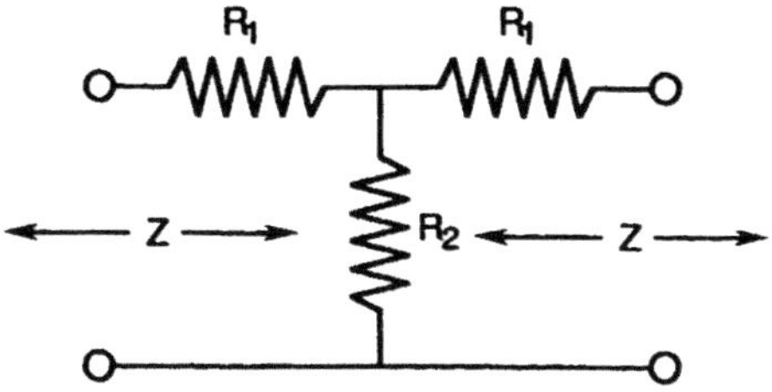

For Z = 100 ohms:

Loss:	R1:	R2:
1 dB	5.74	868
2	11.46	432
3	17.12	284
4	22.6	220
5	28	164.6
6	33.2	134
7	38.2	111.8
8	43	94.6
9	47.6	81.2
10	52	70.4
11	56	60.2
12	59.8	53.6
13	63.4	47.2
14	66.8	41.6
15	70	36.8
16	72.8	31
17	75.2	26.8
18	77.6	24.6
19	79.8	22.4
20	81.8	20.2
21	83.6	18
22	85.2	16
23	87	14.2
24	88.2	12.6
25	89.4	11.3
26	90.4	10
27	91.4	8.9
28	92.2	8
29	93.2	7.1
30	93.8	6.3
35	96.6	3.5
40	98	2
45	99	1.1
50	99	.63
55	99.6	.35
60	100	.2

Figure 109. Design of Symmetrical T-pads

110. DESIGN OF L-PADS

The simpler L-pad can be used in matched impedance systems where it is not necessary for the input of the receiving device to look back into the design impedance, as shown by the arrows in the schematic of Figure 110.

As in the previous section, the design impedance is 100 ohms, and the same conversion method holds.

L-pads are often used in distributed loudspeaker systems for adjusting levels. In this case, it is important that the signal source see the design impedance at all times, whereas the response of the loudspeaker is not primarily dependent upon looking back into the design impedance.

EXAMPLE: Determine values of R_1 and R_2 for an L-pad that will provide 4 dB of signal loss to an 8-ohm loudspeaker.

Determine the values of R_1 and R_2 that correspond to a 4-dB loss. Multiply these values by 8/100 = .08 to arrive at values of 13.7 and 1.8 ohms, respectively. In this case, it is very important to determine the maximum power that will be dissipated by the resistors in the pad and ensure that they have the appropriate power ratings.

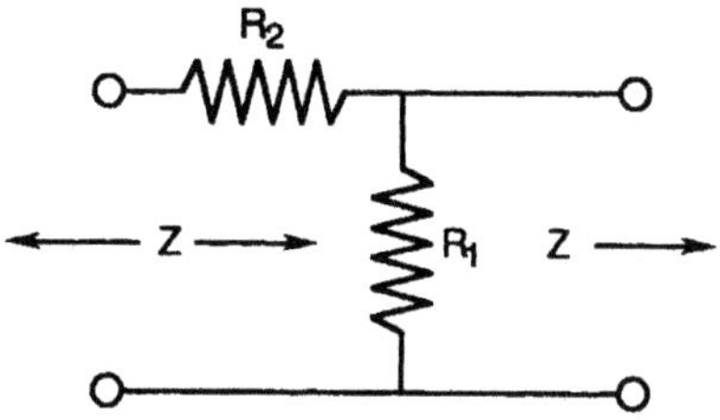

For Z = 100 ohms:

Loss:	R1:	R2:
1 dB	818	5.74
2	385	11.46
3	232.4	17.12
4	171	22.6
5	128.6	28
6	100.4	33.2
7	81	38.2
8	66.2	43
9	54.8	47.6
10	46.2	52
11	39.4	56
12	33.6	59.8
13	28.8	63.4
14	25	66.8
15	21.6	70
16	18.8	72.8
17	16.4	75.2
18	14.4	77.6
19	12.6	79.8
20	11.1	81.8
21	9.8	83.6
22	8.6	85.2
23	7.6	87
24	6.7	88.2
25	6	89.4
26	5.3	90.4
27	4.7	91.4
28	4.1	92.2
29	3.7	93.2
30	3.3	93.8
35	1.8	96.6
40	1	98
45	.56	99
50	.32	99
55	.18	99.6
60	.1	100

Figure 110. Design of L-pads

111. SUMMING OF LEVELS

Levels can be summed by using the graph presented in Figure 111. The sum of two levels is given by the following equation:

$$\text{Level (dB)} = 10 \log[10^{A/10} + 10^{B/10}] \qquad (111.1)$$

where A and B are the two levels being summed.

EXAMPLE: Find the level of the sum of acoustical signal of 100 and 102 dB L_p.

Taking the difference between the two levels, we get 2 dB. Locate 2 dB along the abscissa and read upward to the bold curve. Then find the value of about 2.1 along the ordinate. The value of 2.1 is then added to 102 to arrive at 104.1 dB L_p.

Any number of levels may be summed simply by taking them two at a time and finding the new value.

EXAMPLE: Find the level of the sum of three signals that are all 0 dB.

Taking the first two, we arrive at a value of 3 dB. Then we take the levels of 3 and 0 and find their sum, which is about 1.75 added to 3, or 4.75 dB.

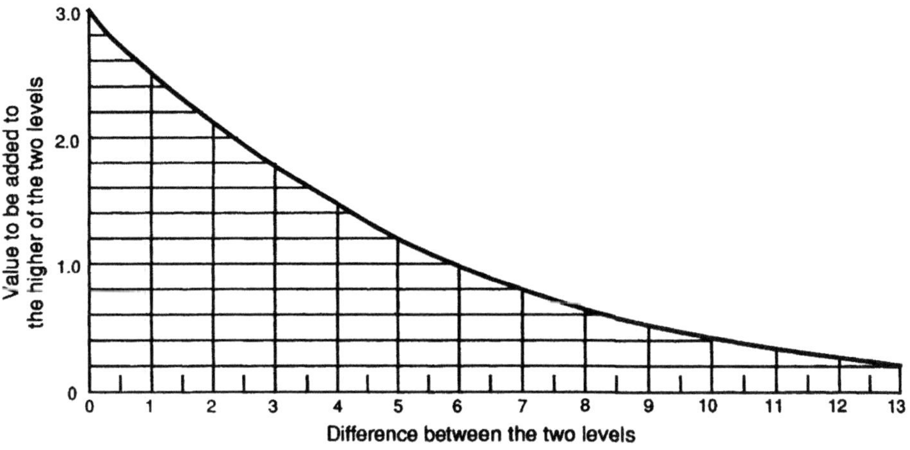

Figure 111. Summing of Levels

112. DISTORTION PERCENTAGE AND LEVEL

Depending on the application, distortion components in a signal can be represented either in percentage or in level below the fundamental. The basic relationship is:

Distortion level (dB) = 20 log(percentage/100) . . . . (112.1)

In Figure 112, the ordinate on the left refers to distortion level in the range from 0 to –60 dB, and the ordinate on the right refers to distortion in the range of –60 to –120 dB.

EXAMPLE: At a specified output, a power amplifier exhibits a second harmonic distortion component that is 50 dB below the fundamental. What is the distortion percentage?

Locate –50 dB along the appropriate abscissa and move upward until the bold line is intersected. Then move to the right ordinate and read the value of 0.3 percent.

EXAMPLE: A microphone produces an open-circuit output of 2 volts in a sound field of 134 dB L_p. The harmonic distortion is measured at 0.5 percent. What distortion level corresponds to this?

On the left ordinate, read downward to 0.5 percent. Then move across to the bold line and read downward to the appropriate abscissa to the value of –46 dB.

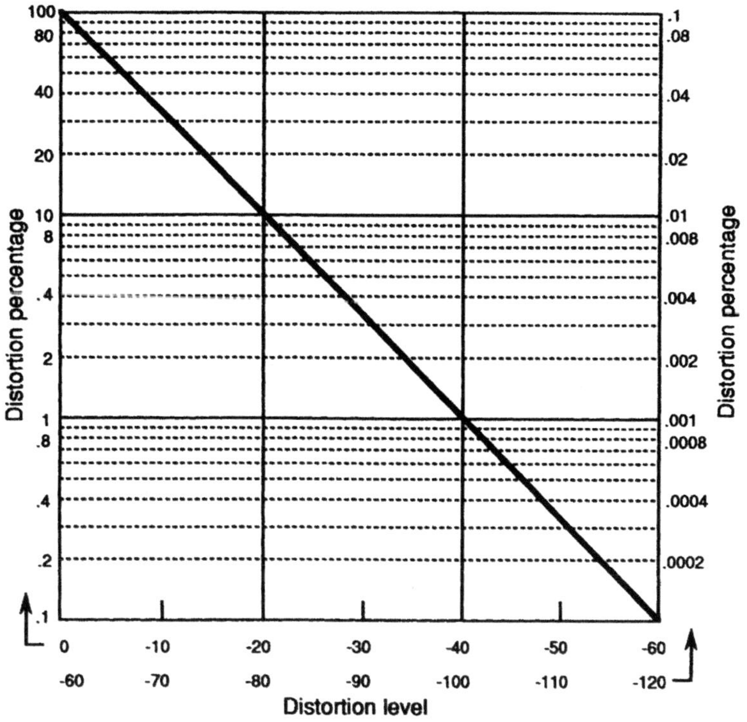

Figure 112. Distortion Percentage and Level

113. Load Impedance as a Function of Power Input in 70-Volt, 100-Volt, and 25-Volt Distribution Systems

Distributed loudspeaker systems often make use of 70-volt or 100-volt distribution methods. In a 70-volt system, the full power of the amplifier, whatever it might be, is always available at 70 volts rms. By means of the appropriate distribution transformers, a given number of watts may be drawn from the line by a given loudspeaker. When the sum of electrical watts taken from the line equals the power rating of the amplifier, the system is fully loaded. The benefit here is that the design engineer does not have to make laborious system load impedance calculations for each loudspeaker placed across the line, but merely has to keep track of the number of watts drawn.

It is useful, however, to know the actual impedance of the fully loaded line so that the proper choice of wiring may be made. Seventy-volt systems are commonly used in the United States. Countries using 220-volt ac power distribution normally use 100-volt systems.

Figure 113 plots the following equations:

For a 70-volt system:

$$Z_{\text{load}} = (70)^2/W_E = 5000/W_E \qquad (113.1)$$

For a 100-volt system:

$$Z_{\text{load}} = (100)^2/W_E = 10{,}000/W_E \qquad (113.2)$$

For a 25-volt system:

$$Z_{\text{load}} = (25)^2/W_E = 625/W_E \qquad (113.3)$$

EXAMPLE: Find the total load impedance in a 70-volt system in which 250 watts of power is distributed.

Referring to Figure 113, locate 250 watts on the abscissa and move upward to the bold line representing 70-volt systems. Then read on the ordinate an approximate value of 20 ohms.

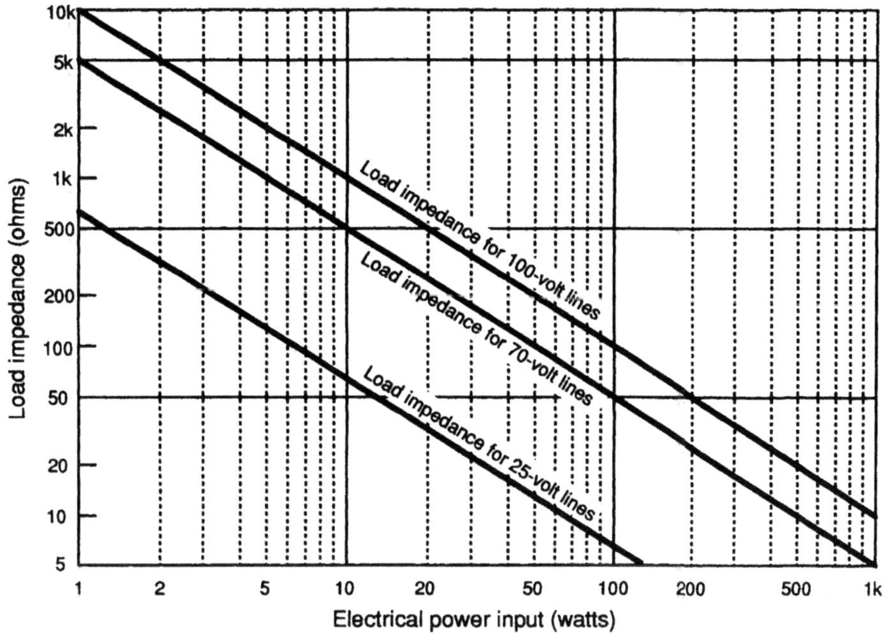

Figure 113. Load Impedance as a Function of Power Input in 70-volt, 100-volt, and 25-volt Distribution Systems

114. Maximum Wire Runs for 0.5 dB Loss in 70-Volt Systems

Modeling a constant-voltage loudspeaker distribution system is rather complex. Calculations can often be simplified by considering all of the load to be placed at the end of the wire run, a worst case consideration. This is shown at Figure 114A and is equivalent to the form shown at Figure 114B.

The tabulation at Figure 114C gives basic data on six wire gauges, expressed both in square millimeter cross-sectional and AWG values. Copper wire is assumed. The right part of the table shows the wire runs for each gauge that will result in no more than a 0.5-dB loss in the line. This is done for three values of load impedance, corresponding to three different powers in the line (see section 113).

EXAMPLE: What is the maximum wire run for a 70-volt distribution system using AWG no. 14 wire and carrying a total power of 200 watts?

First determine the effective load impedance:

$Z = (70)^2/200 = 5000/200 = 25$ ohms

Find the row in the chart for AWG no. 14 wire; then move across to the column for load impedances of 25 ohms. The value is 90 meters (300 feet).

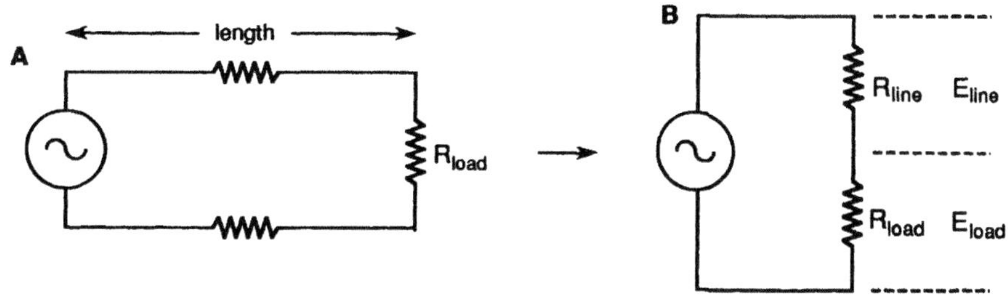

C

Conductor cross-sectional area (mm^2)	AWG no.	Resistance per 300 m (1000 ft) pair	Maximum wire run for 0.5 dB loss — Load impedance 25 ohms:	50 ohms:	100 ohms:
5.2	10	2.00	225 m (750')	450 m (1500')	900 m (3000')
3.3	12	3.15	142 (472)	284 (945)	568 (1890)
2.1	14	5.00	90 (300)	180 (600)	360 (1200)
1.3	16	8.00	56 (188)	113 (375)	226 (750)
0.87	18	12.5	36 (120)	72 (240)	144 (480)
0.52	20	20.0	22.5 (75)	45 (150)	90 (300)

Figure 114. Maximum Wire Runs for 0.5-dB Loss in 70-volt Systems

115. Peak and RMS Values of Waveforms

The nature of a program signal determines the average power that can be delivered by an amplifier. The output voltage waveform can be analyzed and its peak to rms ratio can be determined over fairly short periods of time. The analysis is simplest to make when continuous waveforms are encountered, as shown in this section.

A sine wave is shown at Figure 115A. The peak value of the waveform is arbitrarily set at unity, and the rms value of that waveform, as shown in elementary electrical engineering, is .707. The ratio of peak to rms values of a waveform is known as "crest factor." For the sine wave, it is equal to 1.414, (+3 dB). It is a direct indication of the ratio of peak to average power that an amplifier can deliver when passing a particular waveform.

An amplifier that has a maximum sine wave power rating of 100 watts can actually deliver 200 watts if the signal is a square wave, as shown at Figure 115B. In this case, the crest factor is unity.

The waveform shown at Figure 115C has a crest factor of 1.414 (+3 dB), whereas that shown at Figure 115D has a crest factor of 2 (+6 dB).

Noise waveforms are generally described statistically. They are normally "shaped" by clipping their peak values to create a waveform with a particular crest factor. For example, the waveform shown at Figure 115E has a measured rms voltage value, and a symmetrical clipping is place at the output of the noise generator so that any positive and negative voltages that exceed twice the rms value are clipped. The resulting waveform then has a crest factor of 2 (+6 dB). This particular signal is very useful in testing loudspeakers because its 6 dB crest factor resembles that of highly processed musical program material and its bandwidth extends over a wide range.

The noise waveform shown at Figure 115F has been clipped at positive and negative voltages that are three times the rms value. The crest factor in this case is 3 (+9.6 dB).

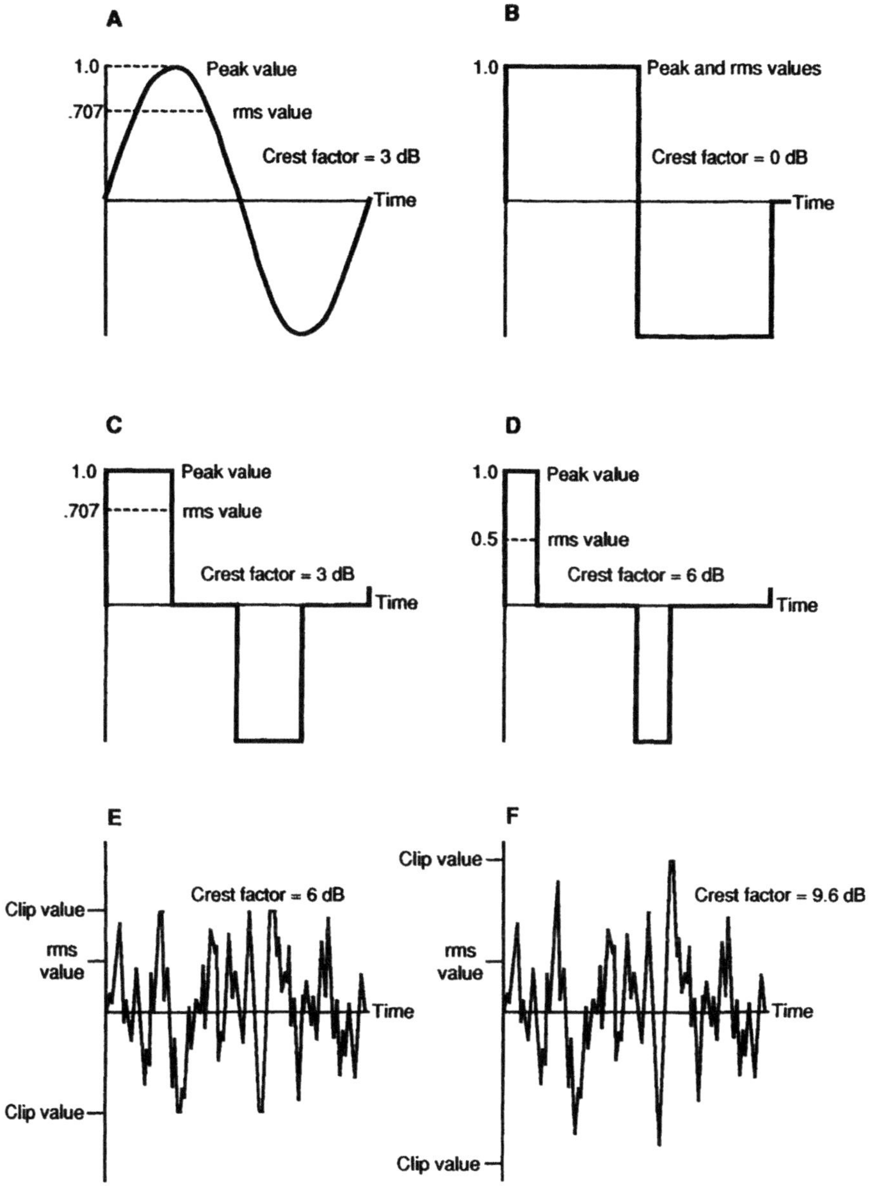

Figure 115. Peak and rms Values of Waveforms

116. Input and Output Impedances of Electronic Devices

Input Impedance

Most consumer electronic devices have fairly high input impedances, normally in the range of 100,000 ohms. Line inputs on today's professional equipment may be in the range of 10,000 ohms, and older professional equipment may have 600-ohm inputs designed for matching. Inputs for phonograph cartridges are often in the 47,000 range, whereas inputs for professional "low impedance" microphones are normally in the range of 2500 to 3500 ohms.

The input impedance of a device can be accurately estimated by observing its loading effect on a signal generator with a source impedance of 600 ohms, as shown at Figure 116A. In this schematic, Voltage A is the unloaded voltage of a signal generator with a 600-ohm output impedance; Voltage B is the output of that generator when it is loaded by the device under test. The difference in these two voltages is calculated in dB. The value in dB is located on the abscissa of the graph shown at Figure 116B. Scanning upward to the bold curve, then across to the ordinate, we read the input impedance directly.

Figure 116B shows a wide range of input impedances, whereas Figure 116C shows a relatively small range of input impedances.

EXAMPLE: We are preparing to make a measurement on a microphone preamplifier. Using the setup shown at A, we observe that Voltage B is 0.084 and Voltage A is 0.1. Find the input impedance of the preamplifier.

First, calculate the level difference between Voltages A and B as −1.5 dB. Referring to the graph, locate −1.5 dB along the abscissa and read up to the bold curve. Move directly across to the ordinate to read the value of approximately 3000 ohms.

Output Impedance

Modern electronic devices normally have fairly low output impedances. Power amplifiers in particular may have output impedances in the range of a small fraction of an ohm. The output impedance of a device can be accurately estimated by removing all loads from it and driving its input with a sine wave at a value that will produce an output in the range of about 1 volt. Then, place a 600-ohm resistor across the device's output and note the voltage drop across it, as shown at Figure 116D.

That value is then expressed in dB and entered into either Figure 116E or F, whichever one gives the greatest resolution. The intersection of the level difference and the bold curve will give the output impedance at the ordinate.

EXAMPLE: An older electronic device is driven with a sine wave at 1 kHz, and the open-circuit output is set at 1 volt. When the device is loaded with a 600-ohm resistor, the output drops to a value of about 0.5 volts. Find the output impedance of the device.

First, calculate the level difference between 1 and 0.5 volts as 6 dB. Then locate –6dB along the abscissa and move upward to the bold line. Then move across to the ordinate to read the value of 600 ohms.

The device under test here is undoubtedly an older model that was normally intended to be used in a studio transmission system in which all input and output impedances were 600 ohms—a so-called matched system.

Reference: Metzler (1993)

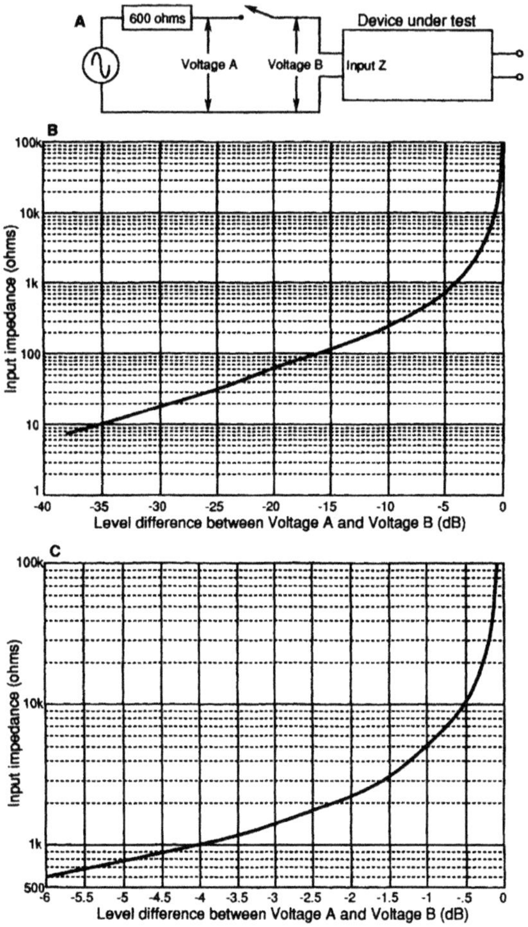

Figure 116. Input and Output Impedances of Electronic Devices

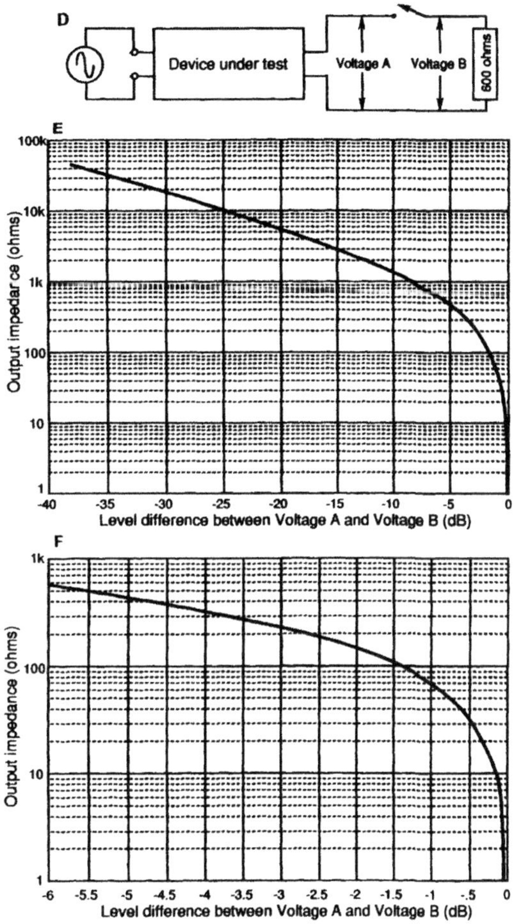

Figure 116. (Continued)

117. Loudspeaker Damping Factor as a Function of Line Length and Wire Gauge

Damping factor is the ratio of loudspeaker impedance divided by the sum of amplifier source resistance and the resistance in the wire connecting the amplifier and loudspeaker. Because the amplifier internal impedance is normally negligible, we consider here only the contribution by the wire, as shown at Figure 117A.

The total wire length is, of course, twice the distance of the loudspeaker from the amplifier.

EXAMPLE: Find the damping factor for an 8-ohm loudspeaker driven by an AWG no. 12 copper wire pair over a distance of 15 meters.

Referring to Figure 117B, locate 15 meters along the abscissa and move upward to the bold line corresponding to AWG no. 12. Then move to the right to the column marked 8 ohms to read the approximate value of 50.

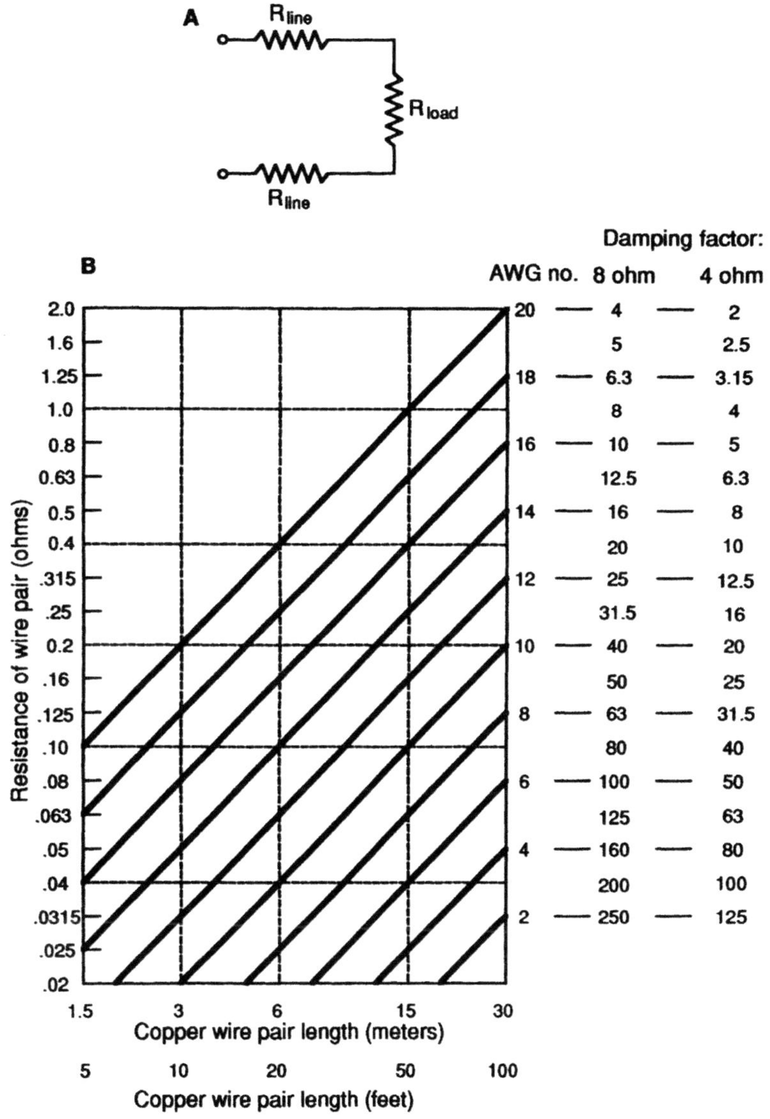

Figure 117. Loudspeaker Damping Factor as a Function of Line Length and Wire Gauge

118. Amplifier Requirements: Direct Field Considerations

The sound pressure level on the axis of a loudspeaker in a free field is given by the following equation:

$$\text{dB } L_p = S + 10 \log (W_E) - 20 \log (r) \quad \ldots\ldots\ldots\ldots\ldots \quad (118.1)$$

where S is the on-axis sensitivity of the loudspeaker measured at 1 meter with 1 watt input, W_E is the electrical power input to the loudspeaker, and r is the measurement distance from the loudspeaker in meters.

There are so many variables in this equation that a single graph does not suffice; we need two graphs. The procedure is as follows: On the ordinate of Figure 118A, locate the desired listening distance from the loudspeaker. Then move upward to the bold line labeled for the desired level. Then move to the ordinate and read the resulting level 1 meter from the loudspeaker.

Moving on to Figure 118B, locate the sensitivity of the loudspeaker along the ordinate. Move to the right to intersect the bold line representing the 1-meter level determined in Figure 118A. Finally, from this intersection, move directly down to the abscissa and read the required power.

EXAMPLE: We want to operate a loudspeaker with a 1-watt, 1-meter sensitivity of 100 dB on-axis at a distance of 10 meters. A level of 94 dB L_p is required. Determine the power required for this.

On Figure 118A, locate the distance of 10 meters along the abscissa and move upward to the bold line marked 94 dB. From that intersection, move to the ordinate and read the value of 114 dB at 1 meter.

On Figure 118B, locate on the ordinate the value of 100 dB for the sensitivity of the loudspeaker. From this point, move across to the bold line representing 114 dB at 1 meter. At that intersection, move downward to the abscissa to read the power requirement of about 25 watts.

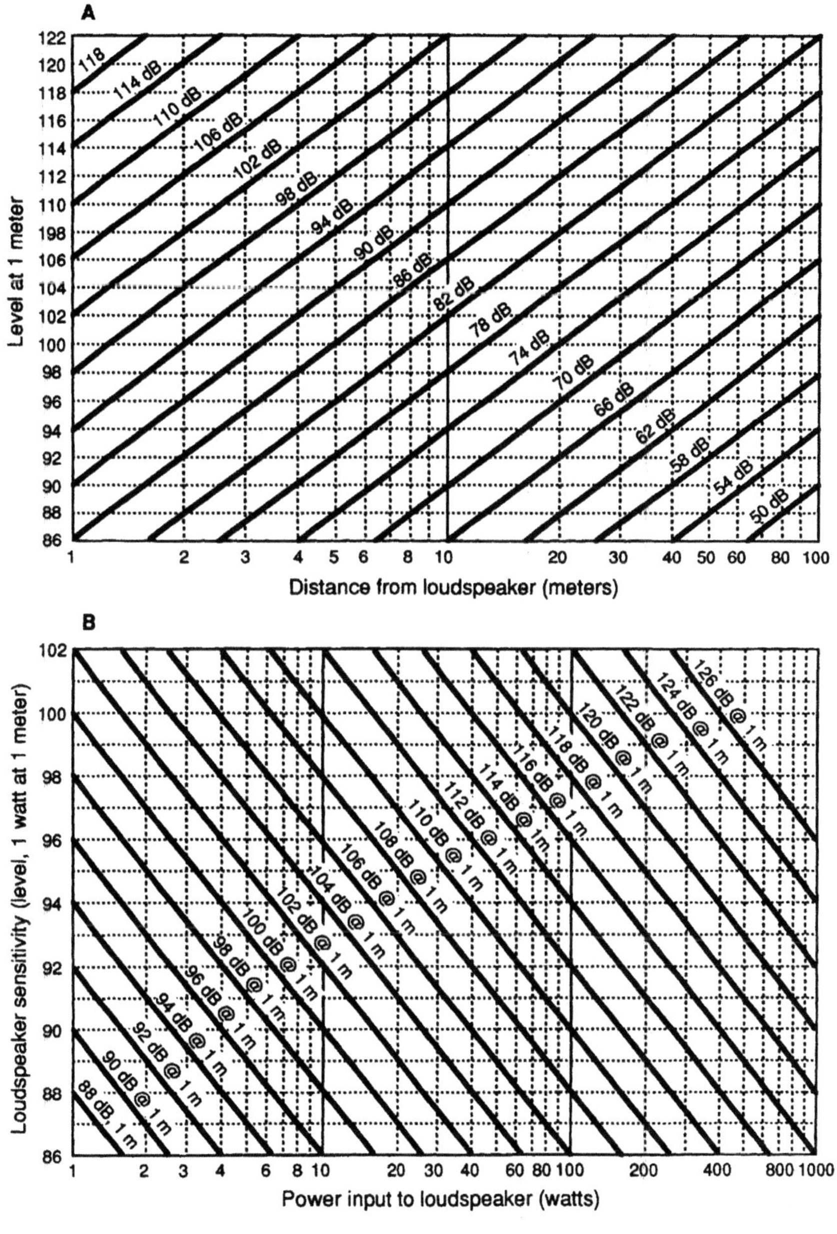

Figure 118. Amplifier Requirements: Direct Field Considerations

119. Amplifier Requirements: Reverberant Field Considerations

There are two principal variables that determine the amount of amplifier power needed to produce a certain sound pressure level in the reverberant field—the total absorption in the room (R, room constant) and the efficiency of the loudspeaker. The pertinent equation here is:

$$L_{reverberant} = 126 - 10 \log R + 10 \log (W_E \times \text{efficiency}) \quad \ldots (119.1)$$

where R is the room constant in square meters and W_E is the electrical power required.

Oftentimes the loudspeaker efficiency is not readily available, but its on-axis directivity index and 1-watt sensitivity are. In this case, the relevant equation is:

$$L_{reverberant} = 17 + 10 \log W_E - 10 \log R - DI + S \quad \ldots\ldots\ldots (119.2)$$

which may be rewritten as:

$$10 \log W_E = 10 \log R + DI + L_{reverberant} - S - 17 \quad \ldots\ldots\ldots (119.3)$$

Many engineers and acousticians prefer to use this equation directly, but the graphical method presented here is sometimes quicker.

EXAMPLE: In a space with a room constant of 10,000 square meters, we want to produce a reverberant level of 100 dB L_p. A loudspeaker with on-axis DI of 8 and 1-watt sensitivity of 105 dB will be used. How much electrical power will be needed?

Referring to Figure 119A, locate the room constant of 10,000 along the abscissa and the desired reverberant level of 100 dB along the ordinate. Determine their intersection at the bold line labeled 25 W_A.

We now know that 25 acoustical watts will be required. Since the efficiency of the loudspeaker has not been provided, we will have to refer to section 46 to relate the DI and S of the loudspeaker to efficiency. Doing this, we arrive at an efficiency of 6.3%.

Moving on to Figure 119B, enter the required acoustical power of 25 W_A along the abscissa and the loudspeaker efficiency of 6.3% along the ordinate. Determine their intersection at a point that is just above the bold line marked 500 W_E. Estimate this at about 400 electrical watts.

As an additional exercise, work out this same problem using equation 119.2.

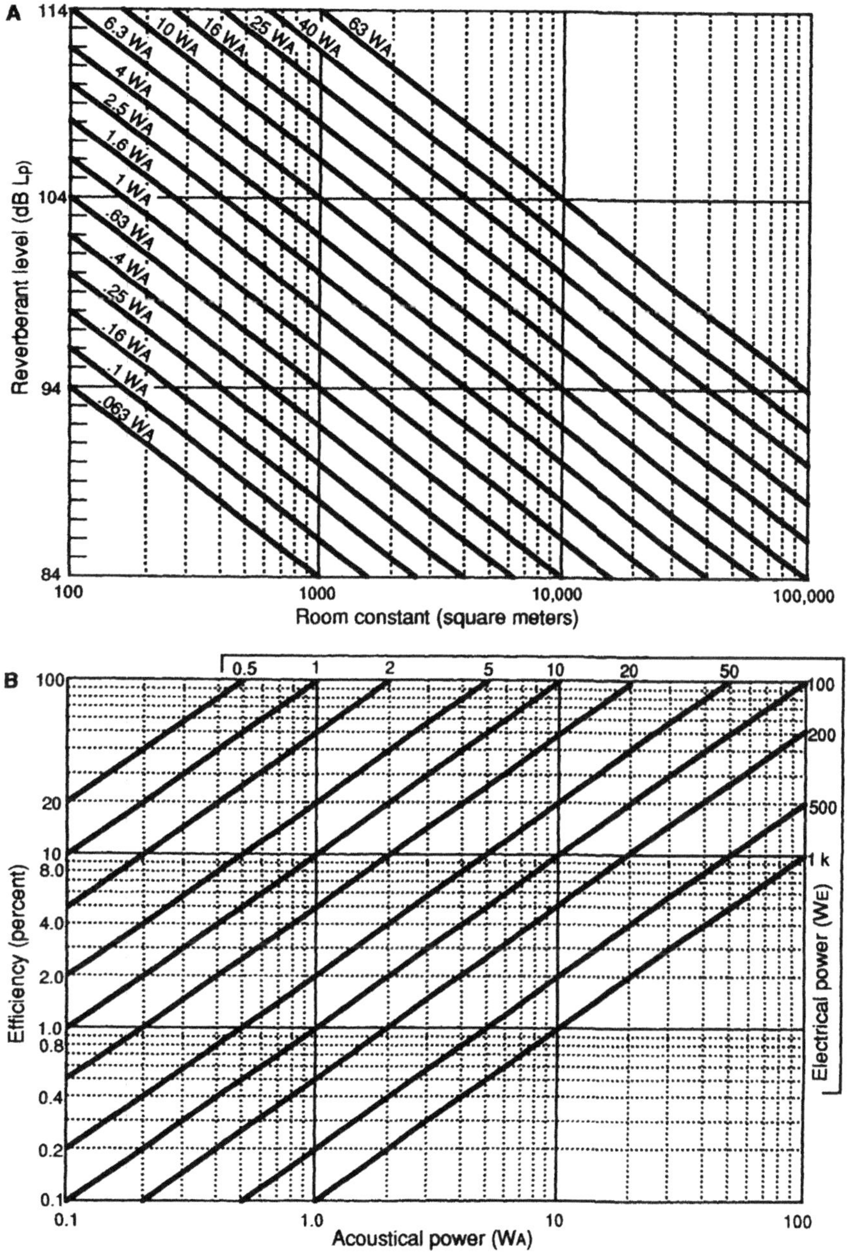

Figure 119. Amplifier Requirements: Reverberant Field Considerations

120. Panpot Response: One Channel to Two

The positioning of monophonic signals on the stereo stage is normally accomplished by panpots (panoramic potentiometers). Figure 120A shows the response of the left and right outputs of a two-channel panpot. Shown at Figure 120B is a corresponding signal flow representation. The angular values of $\pm 90°$ refer to the physical position of the knob on the panpot.

At full left (−90°), there is no output from the right; at the center position (0°), there is equal output at left and right at a reference of 0.7 relative to full output. The corresponding level is −3 dB. At full right (90°), there is no output from the left.

The amplitude transfer functions through the panpot are based on sine and cosine relationships and satisfy the following identity:

$$(\cos \theta)^2 + (\sin \theta)^2 = 1 \quad \ldots\ldots\ldots\ldots \quad (120.1)$$

where θ represents any angle.

If a constant power input signal is panned from left to right, it will remain at constant acoustical power as it makes its transition from the left to the right loudspeaker.

As a practical matter, precise sine and cosine laws do not have to be maintained except over the approximate range from 0 to about −12 dB.

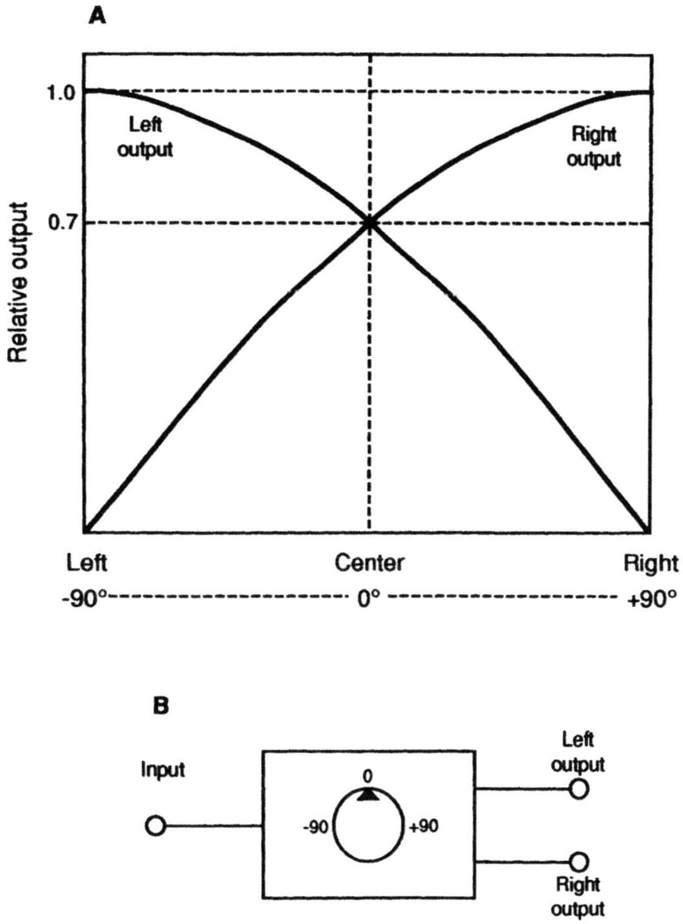

Figure 120. Panpot Response: One Channel to Two

121. Panpot Response: One Channel to Three

Three-channel panpots are widely used in the motion picture industry, where three-screen channels are now standard. The response may be thought of as that of a dual set of two-channel panpots operating only on one adjacent loudspeaker pair at a time.

As in section 120, the angular scale in Figure 121 refers to the physical position of the knob on the panpot.

In some designs, the panpot response does not reach zero at the left, center, and right positions. Some residual leakage is believed to be beneficial for listeners seated at off-center positions in a theater. For example, a signal panned to the left-screen channel can be allowed to leak into the right channel at some low value, perhaps at a level of −20 dB, so that dialogue panned to the left-screen channel is more intelligible to a patron seated at the right front of the theater.

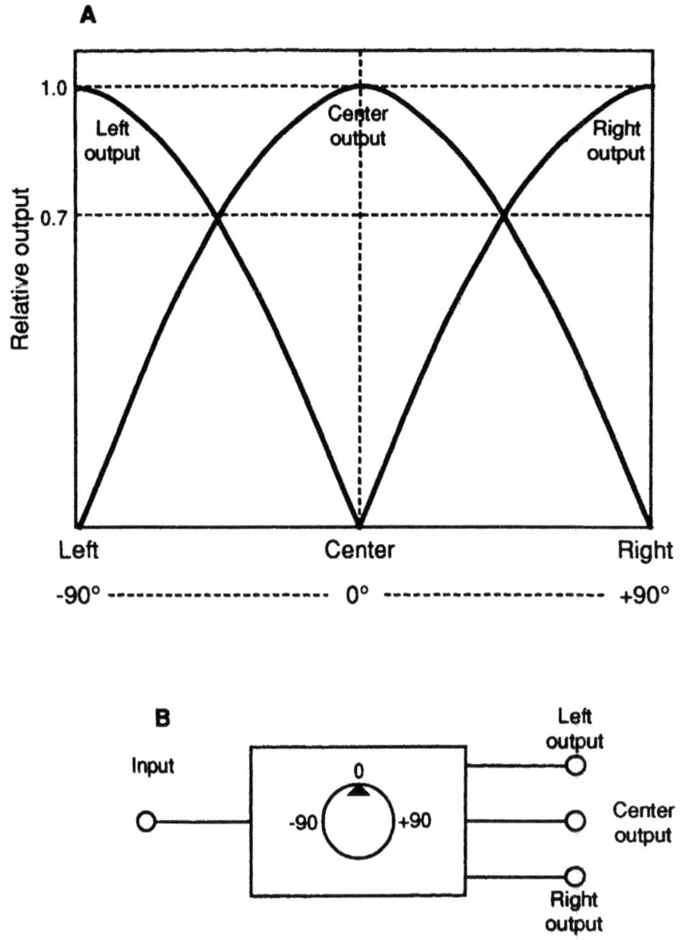

Figure 121. Panpot Response: One Channel to Three

122. Quadraphonic Panpot Response: One Channel to Four

Unlike the panpot response described in sections 120 and 121, the quadraphonic panpot allows the assignment of one input to four simultaneous outputs in a quadraphonic loudspeaker array. Internally, the quad panpot is composed of three one-in/two-out panpots, all coupled together by a joystick arrangement that allows quick assignment of the signals as desired.

As illustrated in Figure 122, one of the panpots steers the signal in the front-back direction, while a pair of panpots steer the signals in the left-right directions.

For example, the input signal may be steered to a single output, say, front-left, by assigning the front-back panpot to front only and the left-right pair to left only. Alternatively, the signal can be steered to all of the output channels simply by positioning all of the panpots at their center positions.

The joystick control facilitates directional assignments in a direct graphical manner; it makes possible almost any conceivable pattern of sound source movement in the azimuthal plane of the loudspeaker array.

As with the panpots discussed in sections 120 and 121, the quadraphonic panpot produces acoustical output power proportional to its input power for all possible positions and combinations.

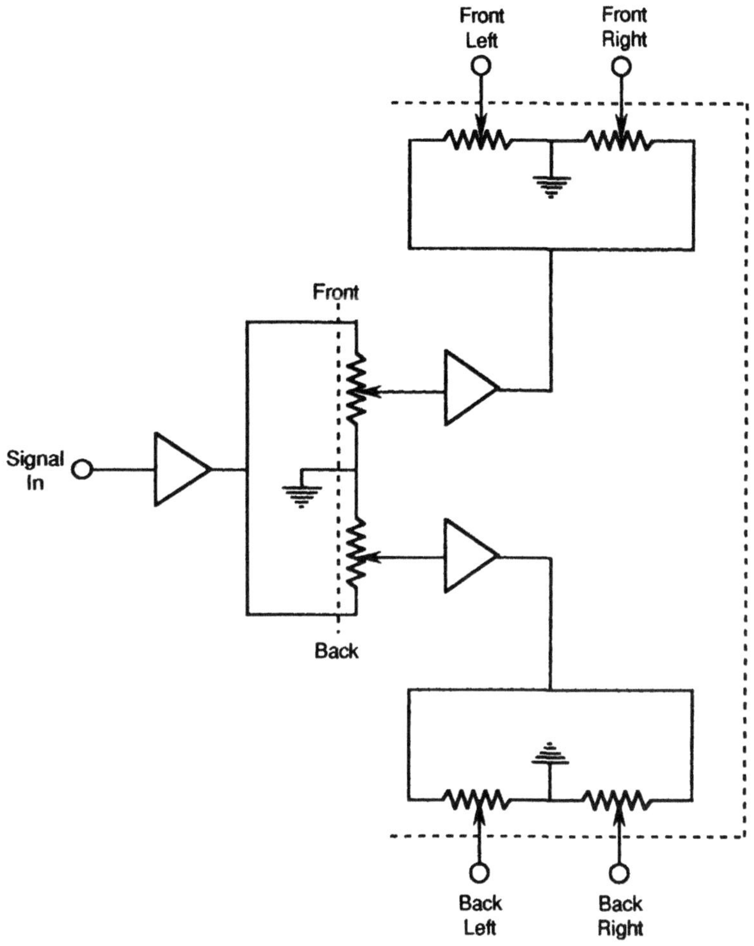

Figure 122. Quadraphonic Panpot Response: One Channel to Four

123. Effect of Noise on Speech Communication

Figure 123 shows the approximate relationship between A-weighted background noise and permissible distances between talkers and listeners for various degrees of communication.

"Satisfactory communication" implies that 95% of all sentences will be understood correctly. A normal voice level at a distance of 5 meters will be clearly understood in noise having a level of 50 dB(A). As the noise level increases, either the talker must raise his or her voice level or the talker-listener distance must be reduced. Generally, for each increase in noise level of 10 dB, the vocal effort will have to increase about 6 dB if the same degree of intelligibility is to be maintained.

For continued voice communication, the talker should never be required to speak any louder than at normal voice, since such prolonged vocal effort is tiresome.

The data given here can be used to determine sound reinforcement acoustical output requirements in noisy environments and workplaces.

Reference: Harris (1979)

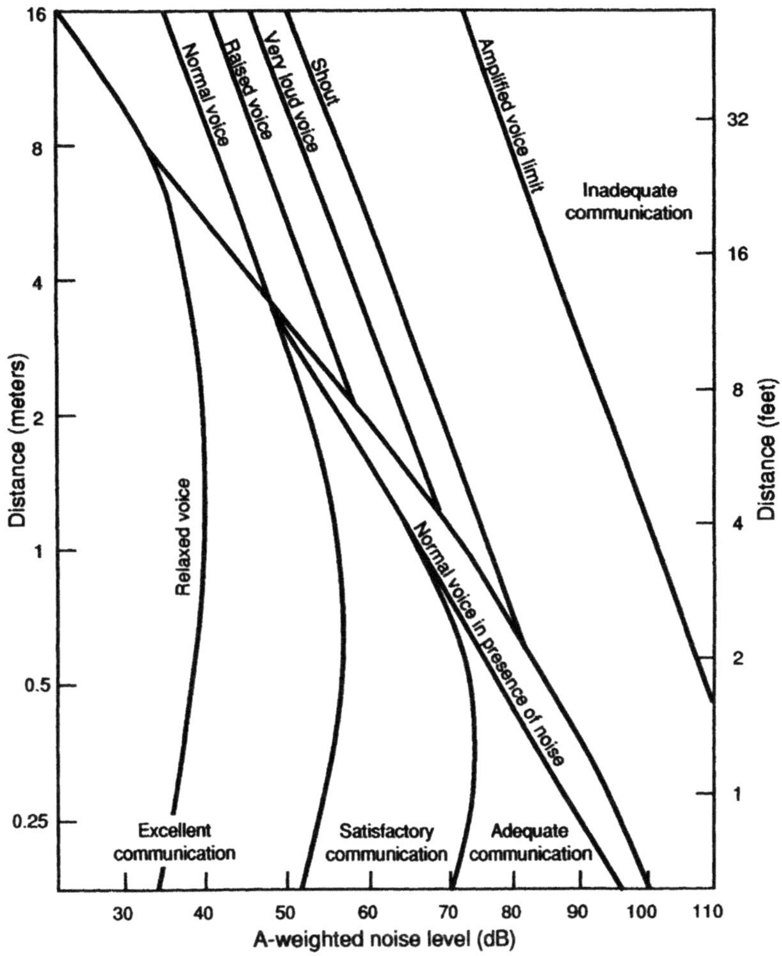

Figure 123. Effect of Noise on Speech Communication

124. Equivalent Acoustic Distance (EAD) and A-Weighted Noise Level

In Figure 124A, D_o represents the distance between listener and talker. Through the use of a speech reinforcement system, that distance can be effectively shortened by raising the level of the talker as perceived by the listener. This has been referred to as "Equivalent Acoustic Distance" or "Expected Acoustic Distance" (EAD).

Figure 124B plots workable speech levels as a function of the noise level in dB(A). The curves indicate raised, normal, and lowered speech levels. They are roughly equivalent, respectively, to levels of 70 dB, 65 dB, and 60 dB as measured 1 meter from the talker.

When speech is heard at low levels, the signal-to-noise ratio should be on the order of about 25 dB. There is a natural tendency for talkers to raise their voices as the noise level increases.

Consider the case of a talker with a raised voice at a distance of 5 meters from the listener. The level of the talker as perceived at that distance will be approximately $70 - 20 \log 5$, or about 56 dB. In a noise environment of 30 dB(A), this will be a signal-to-noise ratio of 26 dB; speech will sound quite clear to a normal listener at a distance of 5 meters.

As the noise level increases, the speech reinforcement system can be adjusted so that it effectively brings the talker closer to the listener, as shown in Figure 124.

If the system is designed for normal speech levels, the middle curve in the graph can be used to determine the EAD. The significance of the dashed portion of the middle curve is that, at elevated noise levels, a person speaking at normal level will tend to raise speech levels to the next higher degree. The same general observations hold for a person speaking in a lowered voice.

In noise levels of 65 dB(A), most people will tend to speak at a raised level, so an EAD of 1 meter may be necessary. In this case, the reinforcement system will produce a speech level at the listener of about 70 dB, with a signal-to-noise ratio of only 5 dB or so. Speech will be intelligible, but both talker and listener will soon tire.

Reference: Eargle (1989)

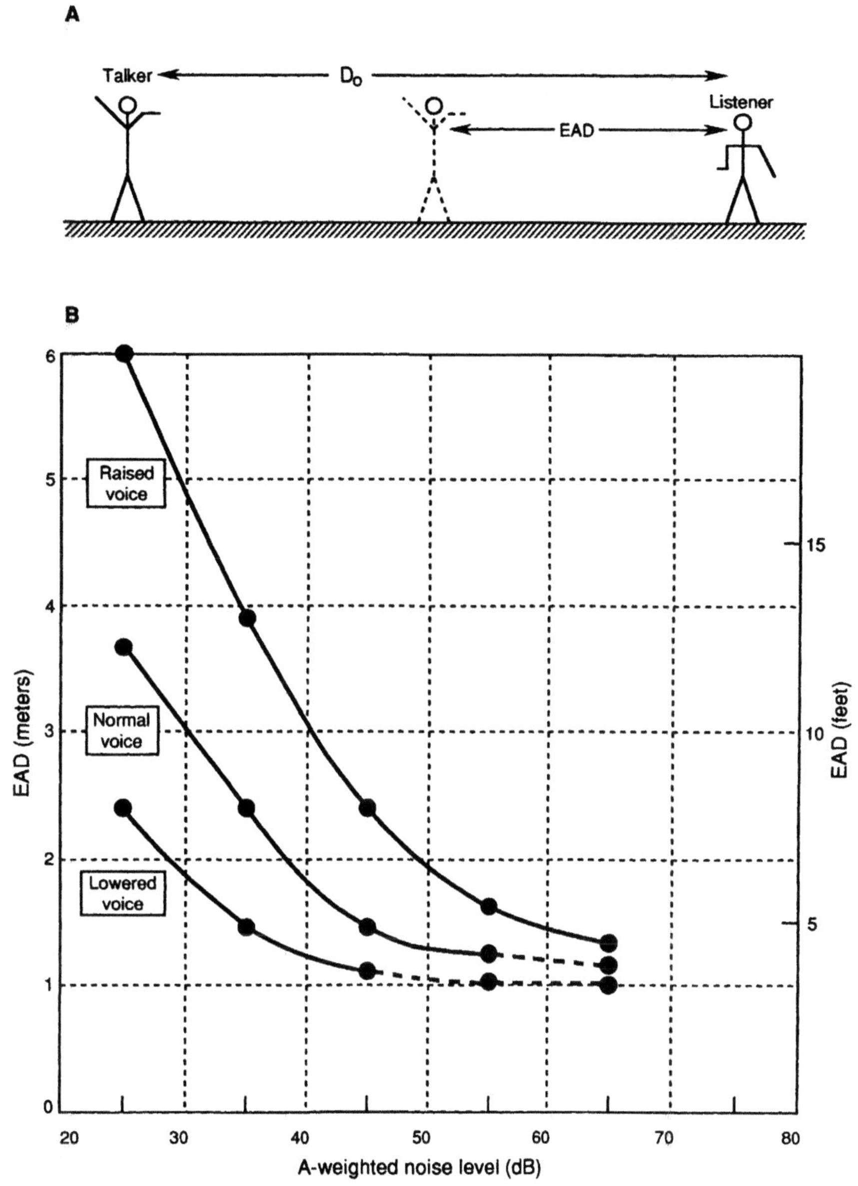

Figure 124. Equivalent Acoustic Distance (EAD) and A-Weighted Noise Level

125. Horn Coverage Angle as Seen in Plan View

The data shown in this section represent an exercise in analytical geometry; they relate the actual coverage angle of a horn to the foreshortened angle as seen in plan view when the horn is tilted downward.

Normally, the useful coverage angle of a horn is determined by the included angle between its –6 dB coverage zones. When a horn with, say, a 90° nominal coverage angle is aimed with a zero elevation angle, its coverage angle as seen in plan view will be 90°. As the angle is tilted downward to an elevation angle of –45°, its foreshortened angle will open up to 110°, as shown at Figure 125A.

The equation that determines the value of the foreshortened angle is:

$$\theta' = 2\arctan[\tan(\theta/2)/\cos\phi] \qquad (125.1)$$

$$(\phi \neq 90^\circ)$$

where θ is the nominal coverage angle of the horn, ϕ is the elevation angle, and θ' is the apparent coverage angle as seen in plan view.

Figure 125B gives values of θ' for selected values of θ at all elevation angles from 0 to 90°.

Reference: Eargle (1989)

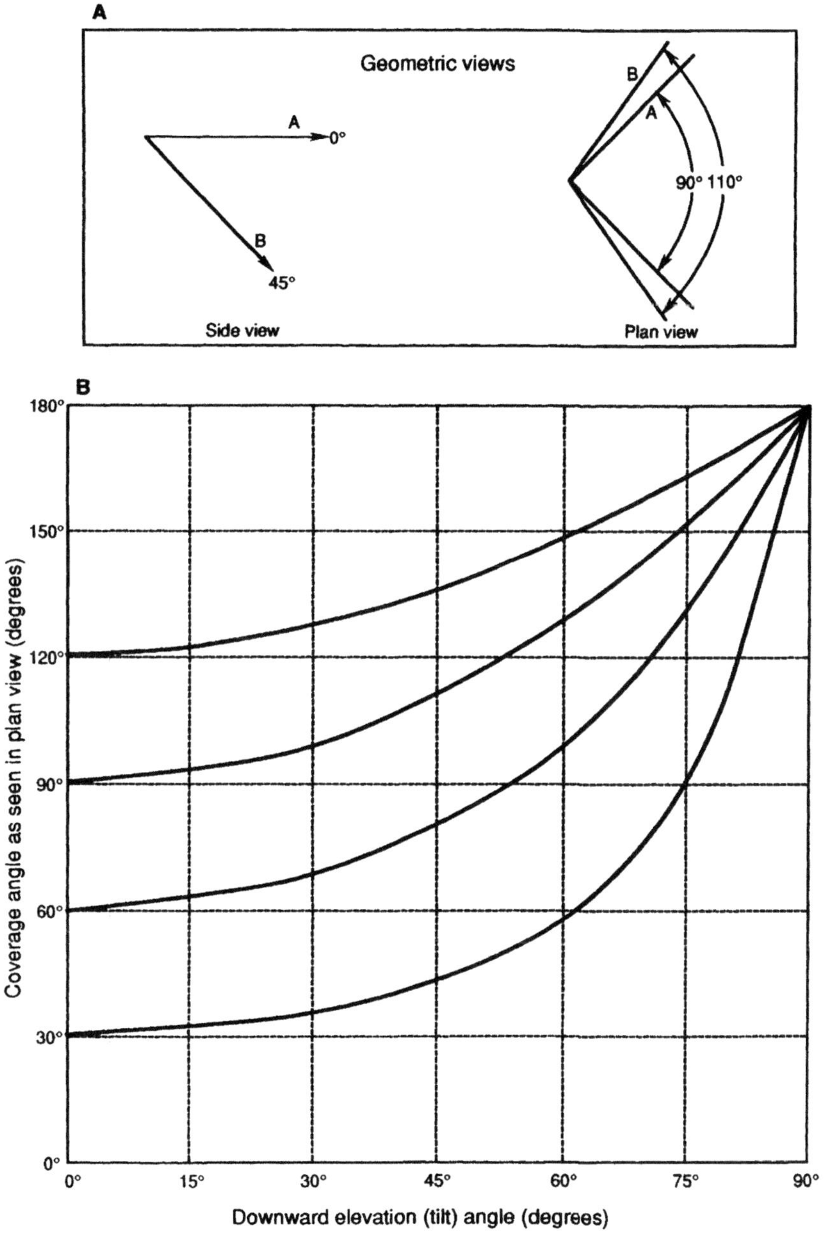

Figure 125. Horn Coverage Angle as Seen in Plan View

126. PEUTZ'S PERCENTAGE ARTICULATION LOSS OF CONSONANTS (Al_{cons})

In its most straightforward form, Peutz's data indicate the percentage of speech articulation loss of consonants by considering the reverberation time in a room and the direct-to-reverberant ratio at 2 kHz as measured at the listener's position. The assumptions are made that the reverberant character of the space follows the general contour given in section 147 and that the signal-to-noise ratio in the space is at least 25 dB. These data are plotted in Figure 126.

Peutz states that this method generally gives results that are within 10% of actual measured values. He further states that an Al_{cons} of 15% represents the maximum loss for acceptable intelligibility.

EXAMPLE: In a room with a reverberation time of 5 seconds and a speech-to-reverberant ratio of 12 dB at the listener, what will be the estimated Al_{cons} value, assuming that noise in the room is at least 25 dB below the speech level?

Locate the value of 5 seconds reverberation time halfway between the bold lines marked 4 and 6. 3 seconds. Find −12 dB along the abscissa and move upward to the 5-second value. Then move to the ordinate and read approximately 45% Al_{cons}. Obviously, speech intelligibility will be very poor.

EXAMPLE: Keeping the same reverberation time of 5 seconds, determine the speech-to-reverberant ratio that will ensure at least 15% Al_{cons}.

Locate the bold curves marked 4 and 6.3 seconds; read between them and find the position corresponding to 15% along the ordinate. Then move down to the abscissa and read the value of approximately −5.5 dB.

EXAMPLE: For a speech-to-reverberant ratio of −12 dB, determine the maximum reverberation time that will allow an Al_{cons} of 15%.

Locate −12 dB along the abscissa and 15% along the ordinate. Determine their intersection point on the graph. At that intersection, read the value of 1.6 seconds from the nearest bold curve.

References: Peutz (1971)
Smith (1981)

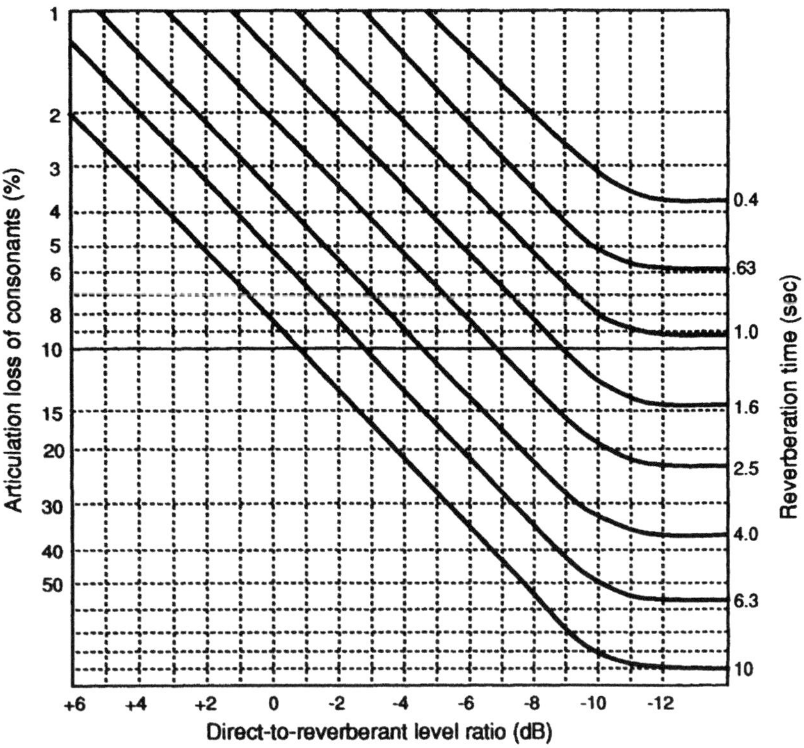

Figure 126. Peutz's Percentage Articulation Loss of Consonants (Alcons)

127. Augspurger's Modification of Peutz's Data

Augspurger replotted that data shown in section 126 as shown in Figure 127. By noting Peutz's 10% accuracy of the method, he divided the graph into four areas: questionable, acceptable, good, and excellent speech intelligibility. The boundaries between the areas correspond to values of 5, 10, and 15% Al_{cons}.

The replotted data are easier to use in speech reinforcement system design, and they give results that are quite close to the original Peutz graph shown in section 126.

Reference: JBL (1982)

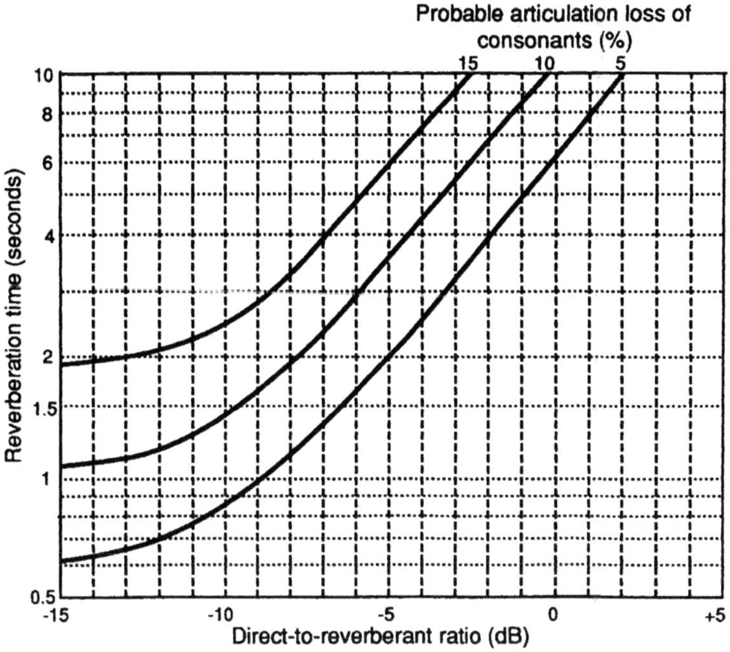

Figure 127. Augspurger's Modification of Peutz's Data

128. CALCULATION OF ARTICULATION INDEX (AI)

The Articulation Index was formulated by French and Steinberg as a means of quantifying speech intelligibility over telephone transmission systems where noise, rather than reverberation, was the primary cause of intelligibility loss. Later modifications were made by Kryter. AI is a multiband evaluation based on signal-to-noise measurements in the following octave bands: 250, 500, 1000, 2000, and 4000 Hz.

The 2000-Hz band is weighted most strongly; the 4000- and 1000-Hz bands are given somewhat less weight, and the 500- and 250-Hz bands are given the least weight.

The individual octave band signal-to-noise ratios are shown at Figure 128A. AI coefficients are read along the right ordinate. The AI is the sum of all five coefficients.

EXAMPLE: For a given space, the rms background noise spectrum at a given listening position is as shown at Figure 128B. A speech reinforcement system provides a long-term amplified spectrum, as shown for male announcers. Determine the articulation index at the given listening position.

Plot the noise and signal spectra on the same graph and note the signal-to-noise ratio at each octave band. Beginning with the 250 Hz, enter the signal-to-noise value along the ordinate of the graph shown at Figure 128A and read upward to the bold line marked 250 Hz. Then read across to the ordinate at the right. Note the coefficient value of approximate 0.03. Do the same for the remaining four octave bands; then sum them to get the AI value of 0.59.

An AI value of unity (background noise 18 dB below speech) is the maximum value. Values in the range of 0.4 and 0.5 are about the lowest that can ensure good intelligibility of speech when heard in sentence context.

Note that both speech and noise spectra are long-term or average values. Peak values of speech will be some 12 dB greater than average values.

References: Kryter (1962)
French and Steinberg (1947)
Smith (1981)

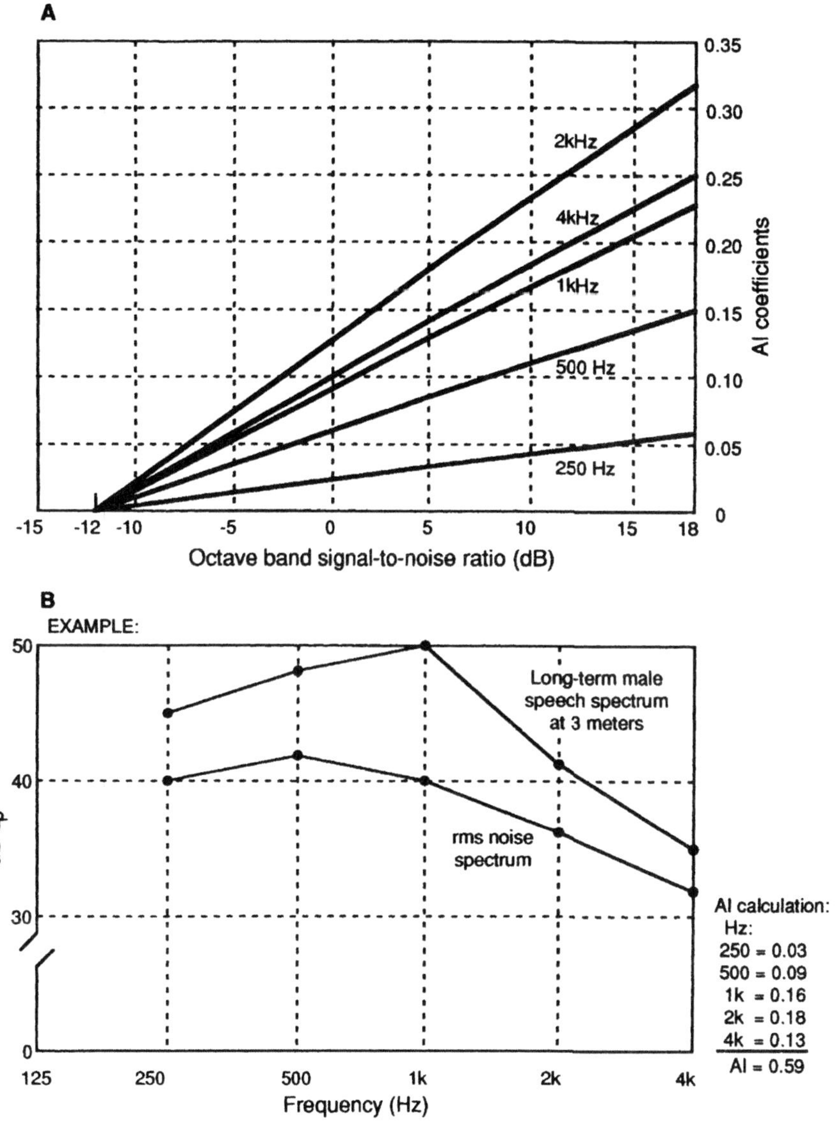

Figure 128. Calculation of Articulation Index (AI)

129. Typical Motion Picture Screen Losses

The data presented here are based on on-axis and off-axis response measurements of a uniform coverage high-frequency horn with and without a screen. The vinyl screen used had 8% area perforation, and the thickness was approximately 0.5 mm. All response curves have been normalized to the 0° incidence curve without the screen.

The data shown at Figure 129A are fairly normal for off-axis response of a nominal 90° × 40° uniform coverage horn that would normally be used in speech reinforcement systems.

When that horn is placed behind a screen and fires into it normally, the measured data are as shown at Figure 129B. Along the primary axis, the screen introduces a loss that is approximately 6 dB per octave, commencing at about 4 kHz.

However, along the various off-axis angles, the response above 6300 Hz may be greater than on-axis. The effect here is not well understood, but it may have to do with the coincidence effect in which transverse vibrations in the screen are in step with the corresponding vector component of off-axis radiation from the horn. Under this condition, the screen will offer very little impedance to the transmission of sound.

Some thinner screen materials will show no on-axis loss below about 8 kHz, thus requiring less electrical equalization of the loudspeakers behind the screen.

The data given here are incomplete. They are presented only for the broadest tutorial purposes. In general, the on-axis loss through the screen should be accounted for in electrical equalization of the signal, while the unusual off-axis phenomena can be ignored.

Reference: Eargle, Bonner, and Ross (1985)

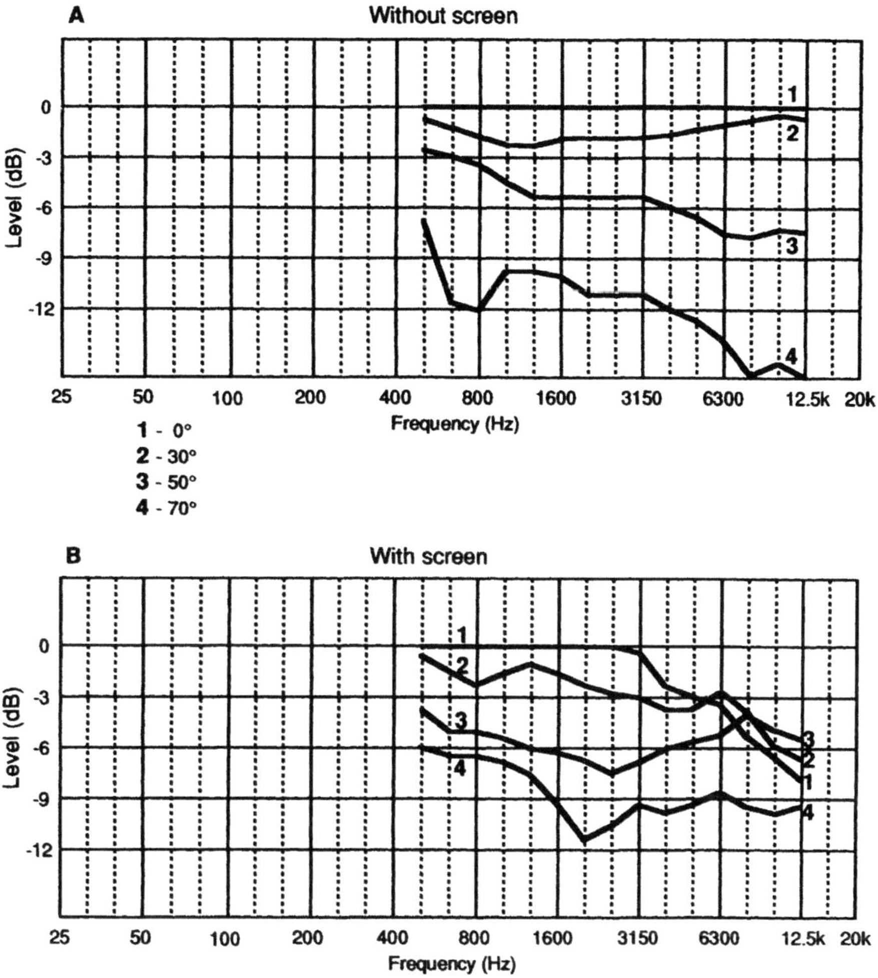

Figure 129. Typical Motion Picture Screen Losses

130. HOUSE EQUALIZATION STANDARD FOR MOTION PICTURE SYSTEMS

The frequency response contour shown in Figure 130 is specified in ISO document 2969. It represents the response of each screen channel as averaged in the house, normally at four positions.

The tolerance on the curve is ±2 dB above 80 Hz. Below 80 Hz, the response may vary in the negative region only, depending on the type of loudspeakers.

In the range from 80 down to 30 Hz, subwoofers are used in premier houses for bass extension of special effects. These devices are normally operated at increased levels to compensate for loudness contours, as discussed in sections 133–135.

In many smaller houses and studio dubbing theaters, the high-frequency response is carried out to 16 kHz.

Reference: Dolby Laboratories (1993)

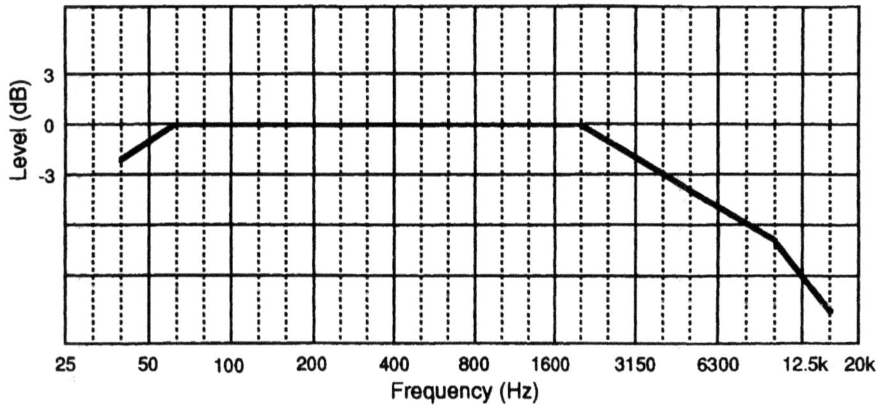

Figure 130. House Equalization Standard for Motion Picture Systems

131. House Equalization for Motion Picture Systems: Adjustments for House Size

Standard practice in the motion picture industry calls for adjusting the basic ISO curve for small and very large theaters, as shown in Figure 131. An adjustment at 10 kHz for small theaters results in a 2-dB boost over the normal ISO standard, whereas an adjustment for very large houses results in a reduction of 2 dB at 10 kHz.

These adjustments are largely empirical, but they do have a basis in experiments where subjective judgments have shown that, in larger rooms, a flat response near-field loudspeaker is best matched by a distant loudspeaker that has a slightly rolled-off high-frequency response in steady-state measurements.

In a large room, the reverberant field is more significant than in a small one; this effect tends to produce greater high-frequency sound pressure in the ear canal than would be felt in a smaller room.

References: Dolby Laboratories (1993)
Holman (1988)
Staffeldt (1984)

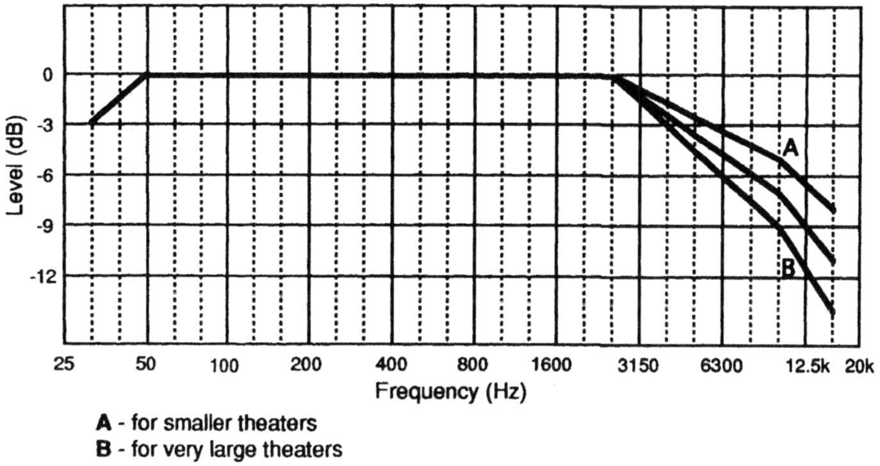

Figure 131. House Equalization for Motion Picture Systems: Adjustments for House Size

132. ISO (International Standards Organization) Preferred Numbers

The ISO preferred numbers are a logarithmic series of numbers based on the 40th root of 10, which is equal to 1.0592537. In Figure 132, in the column labeled 1/12-Octave Series, each successively higher number is produced by multiplying the preceding one by the 40th root of 10. After 40 such multiplications, we reach the number 10; for the numerical decade of 1 to 10, there will be 40 logarithmically spaced intervals. The irrational values of the roots are normally rounded off as shown.

Usually, we speak of the ISO numbers in terms of the number of intervals per octave rather than per decade. This is generally more convenient; it is based on the fact that the 12th root of 2 is equal to 1.0594631.

Thus, the one-twelfth octave series corresponds to the 40th root of 10, the one-sixth octave series to the 20th root of 10, the one-third octave series to the 10th root of 10, and the half-octave series to the sixth root of 10.

The one-third octave series is encountered most commonly in equalizer design and in acoustical measurements.

1/12-Octave Series	1/6-Octave Series	1/3-Octave Series	1/2-Octave Series	1-Octave Series
1.0	1.0	1.0	1.0	1.0
1.06				
1.12	1.12			
1.18				
1.25	1.25	1.25		
1.32				
1.4	1.4		1.4	
1.5				
1.6	1.6	1.6		
1.7				
1.8	1.8			
1.9				
2.0	2.0	2.0	2.0	2.0
2.12				
2.24	2.24			
2.36				
2.5	2.5	2.5		
2.65				
2.8	2.8		2.8	
3.0				
3.15	3.15	3.15		
3.35				
3.55	3.55			
3.75				
4.0	4.0	4.0	4.0	4.0
4.25				
4.5	4.5			
4.75				
5.0	5.0	5.0		
5.3				
5.6	5.6		5.6	
6.0				
6.3	6.3	6.3		
6.7				
7.1	7.1			
7.5				
8.0	8.0	8.0	8.0	8.0
8.5				
9.0	9.0			
9.5				
10.0	10.0	10.0		

Figure 132. ISO Preferred Numbers

PART FIVE

PSYCHOACOUSTICAL DATA

133. Fletcher-Munson Equal Loudness Contours

The Fletcher-Munson equal loudness contours, shown in Figure 133, were derived from measurements made on many subjects during the 1930s. The data relate the subjective loudness of tones as compared with a fixed level at 1000 Hz. Listening was by way of headphones.

For example, a 1000-Hz tone presented to the listener at 50 dB L_p is established as a reference, and a tone at 100 Hz is adjusted by the listener so that it sounds "just as loud" as the reference tone at 1000 Hz. An average of many judgments here indicates that the 100-Hz tone must be about 67 dB L_p in order to sound as loud as the 1000-Hz reference tone. Other tones are compared with the 1000-Hz reference, and a contour is established over the frequency band. That contour is called the "50-phon loudness contour." Note that only at 1000 Hz are the loudness levels in phons equal to the L_p values.

Similar judgments are recorded for other 1000-Hz reference levels, and a family of equal loudness contours is established.

The most striking aspect of these data is that, as very low levels, bass frequencies must be raised considerably if they are to be as subjectively loud as higher frequencies. For example, at a 10-dB L_p reference at 1000 Hz, a 100-Hz tone must be raised some 34 dB if it is to sound as loud as the 1000-Hz tone.

Reference: Fletcher and Munson (1933)

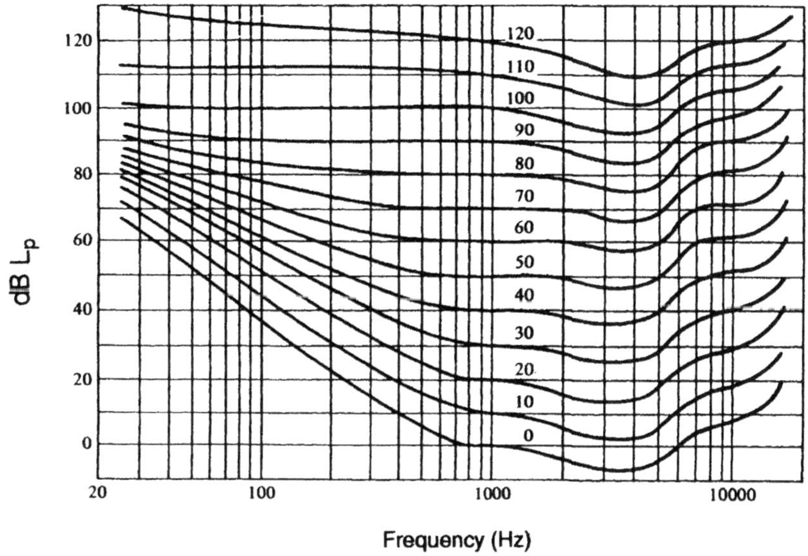

Figure 133. Fletcher-Munson Equal Loudness Contours

134. Robinson-Dadson Equal Loudness Contours

The Robinson-Dadson contours, shown in Figure 134, were established during the 1950s and have largely superseded the earlier Fletcher-Munson data. The following conditions apply to the data:

- The source of sound is directly ahead of the listener.
- The sound reaching the listener is essentially a free plane wave.
- The sound pressure is measured in the absence of the listener.
- Listening is binaural.
- The population of listeners have normal hearing and are in the age group from 18 to 25 years.

The major difference between the Fletcher-Munson data and the Robinson-Dadson data is to be seen in the region below 1000 Hz and below about 90 phons. The differences above about 4000 Hz are fairly constant between the two sets of data and probably have to do with the fundamental differences between headphone and loudspeaker listening.

By comparison, at a 10-dB L_p reference at 1000 Hz, a 100-Hz tone must be raised some 19 dB if it is to sound as loud as the 1000-Hz tone. The difference is substantial.

Reference: Robinson and Dadson (1956)

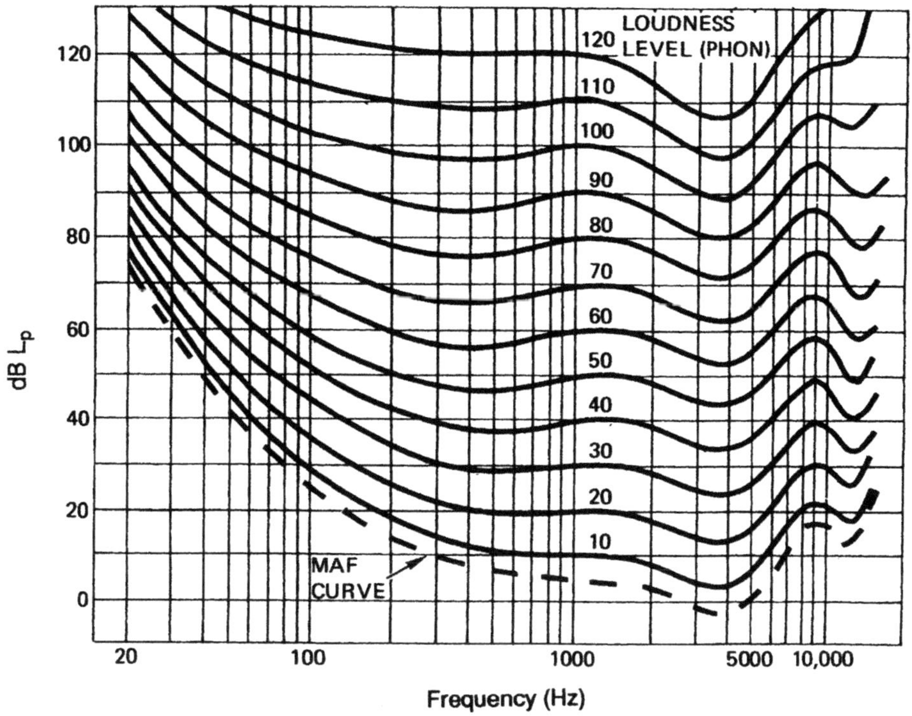

Figure 134. Robinson-Dadson Equal Loudness Contours

135. Churcher-King Equal Loudness Contours

The Churcher and King curves, shown in Figure 135, were based on pure tones and date from 1937. The resemblance to the Fletcher-Munson data is evident, and we presume the tests were made via headphones.

The dashed curves superimposed on the Churcher and King data were generated by Pollack and are based on bands of noise 250 mels wide (see section 142).

As can be seen in this section and the two preceding ones. The judgment of equal loudness depends both on the presentation of the stimulus (headphones versus loudspeaker) and the nature of the stimulus (tones versus noise). With musical signals, it is easy to assume that only a general assessment of equal loudness phenomena may be made and that the broadest of guidelines need to be followed in equalizing music for playback at levels higher or lower than perceived at the original performance.

References: Beranek (1954)
Churcher and King (1937)
Pollack (1952)

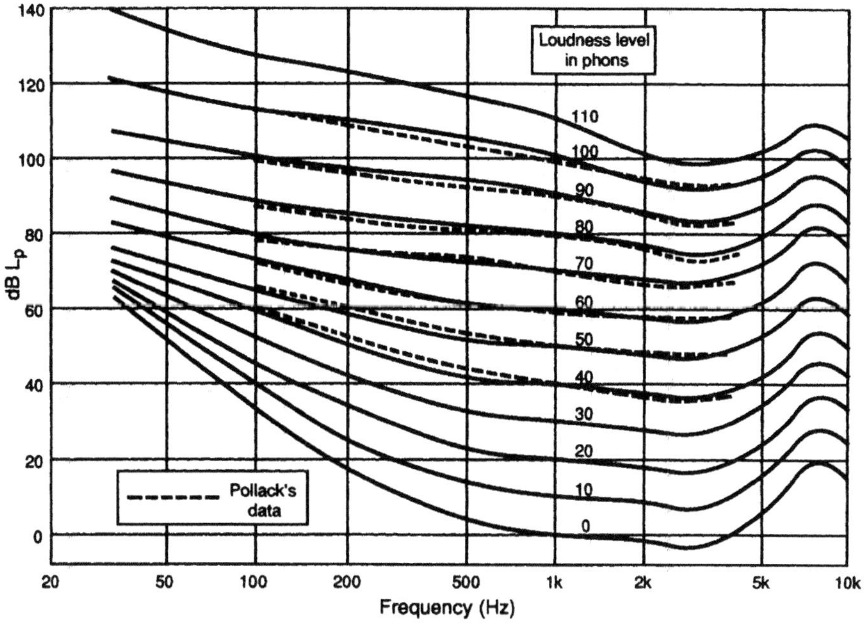

Figure 135. Churcher-King Equal Loudness Contours

136. Determination of "Twice Loudness" at Low Frequencies

At frequencies above about 300 Hz, a level difference of 10 dB corresponds to a subjective judgment of "twice as loud" or "half as loud." Note in the data of section 134 that the 10-dB loudness contours do in fact remain spaced by 10-dB over this range. However, at low frequencies and at lower levels, the 10-dB phon contours converge.

The data of section 134 have been replotted in Figure 136 to show this convergence for phon levels in the range of 90–100 and 40–50. In the 40- to 50-phon range, a level difference of 5 dB will be judged to be twice or half as loud, whereas at 20 Hz a level difference of 2.5 dB will suffice for this same judgment.

In practical terms, this means that music spectral balance must be carefully established if proper balances are to be maintained at low listening levels. Just a slight increase in overall level can suffice to give the impression of too much bass.

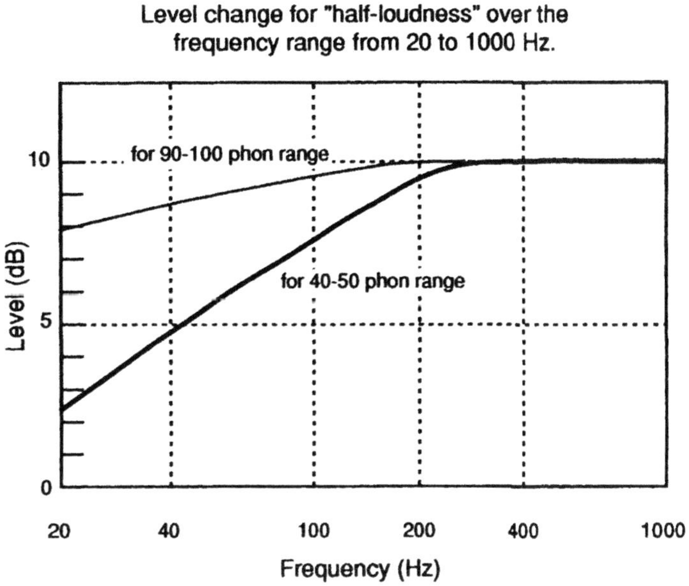

Figure 136. Determination of "Twice Loudness" at Low Frequencies

137. Calculation of Loudness in Sones

Loudness level is measured in phons, whereas loudness is measured in sones. The scale shown in Figure 137A shows the relationship between phons and sones for pure tones. Note that a loudness level of 40 dB is equivalent to 1 sone. Successive increments of 10 dB in loudness level correspond to doubling the value in sones.

For single frequencies, we can read the phon level from section 134 and relate it directly to loudness in sones from the scale at Figure 137A. But for complex signals, the determination of loudness is more complicated. Under this condition, loudness is defined as:

$$\text{Loudness} = S_{max} + 0.3(\Sigma S - S_{max}) \quad \text{(137.1)}$$

where S_{max} is the loudness of the loudest octave band and ΣS is the sum of all octave band loudness indices.

An example is shown in Figure 137B. Individual levels in each octave band from 63 Hz to 12.5 kHz are entered into the chart as shown. The loudest band is at 4 kHz; it has a loudness index of 20. The loudness indices of the remaining octave bands are determined and added together. This value is multiplied by 0.3 and added to the loudness index of the loudest bank. In this example, the resulting loudness is 41.6 sones.

Loudness is normally determined as a calculation and not as a direct measurement. Loudness meters have been developed, but they are cumbersome and rarely used. As a rule, acousticians use weighting curves in conjunction with sound level meters in order to get a first approximation of loudness level. This is the subject of the next section.

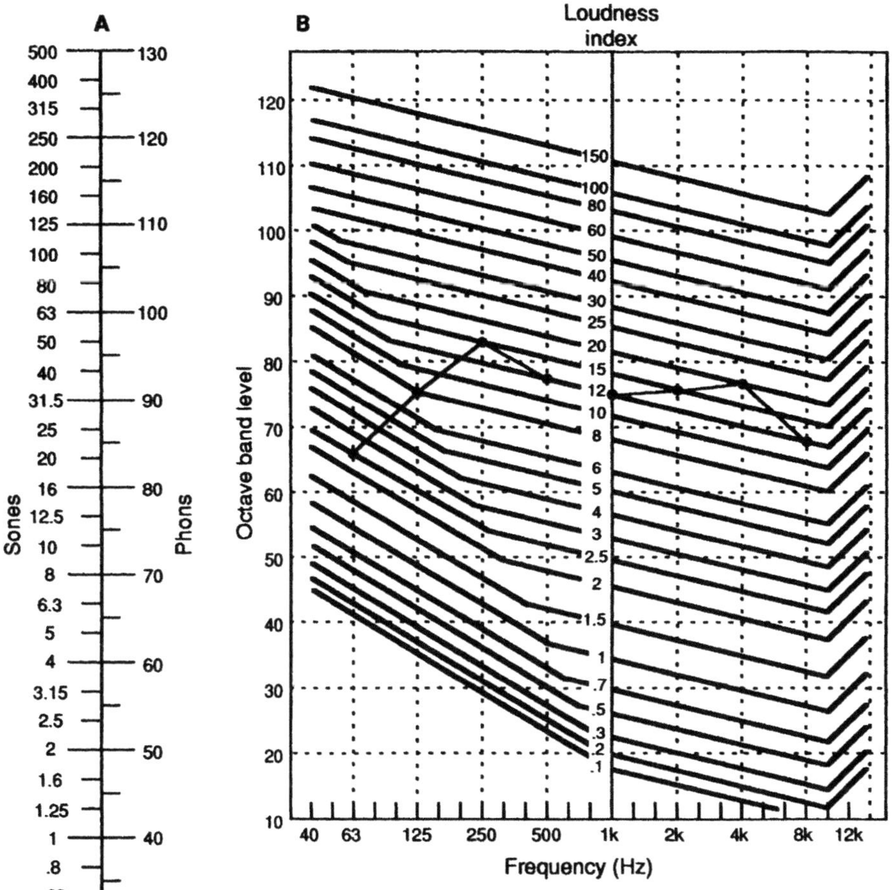

A sample loudness calculation for octave noise bands:

1. Loudest band: 4 kHz. Index = 20
2. Sum of all indices:

Band	63	125	250	500	1k	2k	4k	8k
Index	2.5	8	15	12	12	15	20	10

Index total = 92; therefore, loudness = 20 + 0.3(92 - 20) = 41.6 sones

(*Note:* For 1/3-octave bands, use a multiplier of 0.15.)

Figure 137. Calculation of Loudness in Sones

138. Standard Weighting Curves

Sound level meters normally employ a set of weighting curves so that the meter's readings can be more closely correlated with subjective loudness. In addition, there is normally a flat scale for making absolute measurements of sound pressure level.

Readings on the A, B, and C scales in Figure 138 are normally indicated as dB(A), dB(B), and dB(C). These three weighting curves are roughly the inverse of the 100-, 70-, and 40-phon curves, as seen in the previous sections dealing with equal loudness contours.

It follows logically that the A scale should be used in assessing the overall loudness of low-level noise, inasmuch as the ear is so much less sensitive to low-frequency noise under that condition.

D weighting is found on some sound level meters, but it is rarely used. It was developed for use in aircraft measurements, but most acousticians find the A scale adequate for this purpose.

References: Bruël and Kjær (1971)
Harris (1979)

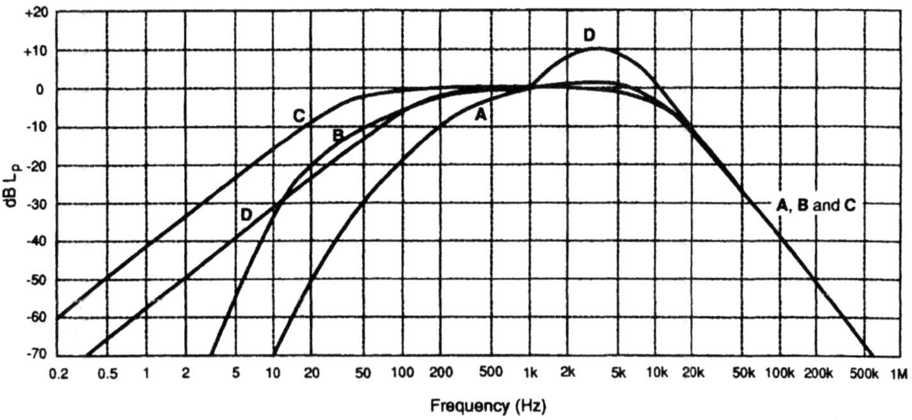

Figure 138. Standard Weighting Curves

139. Loudness and Signal Duration

Tones of short duration at a fixed level will not sound as loud as tones of longer duration at the same level. The effect is observed up to a duration of about 200 to 300 msec, beyond which the loudness will remain the same. The general effect is shown in Figure 139.

It is not by chance alone that the venerable VU meter reaches its full-scale reading in about 300 msec. The device has had a long history in speech transmission as a general indicator of loudness, primarily because its ballistics track the data shown in this section. See section 100 for added discussion.

In general, a decade relationship in duration can be approximately offset by a 10-dB difference in level.

Reference: Watson and Gengel (1969)

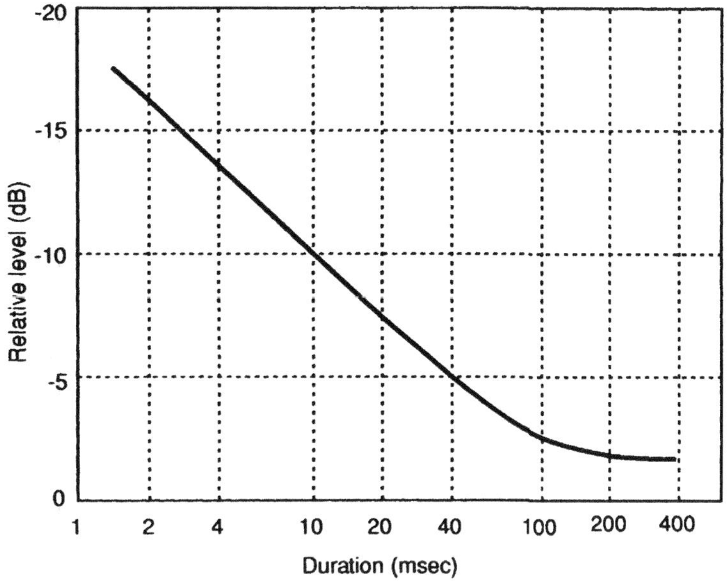

Figure 139. Loudness and Signal Duration

140. Pitch and Level Relationships, I

For pure tones, there are shifts in perceived pitch that are due to loudness. These shifts are smaller than earlier experiments indicated; however, general trends remain the same. Frequencies above 2 kHz generally rise in pitch with increasing level, whereas those below 2 kHz generally fall with increasing level. The data presented in Figure 140 were developed by Terhardt and are based on 15 subjects.

In general, it is difficult to detect pitch changes in complex tones as a function of level, but they can be detected during the reverberation that often accompanies organ music heard in large spaces.

References: Rossing (1990)
Terhardt (1979)

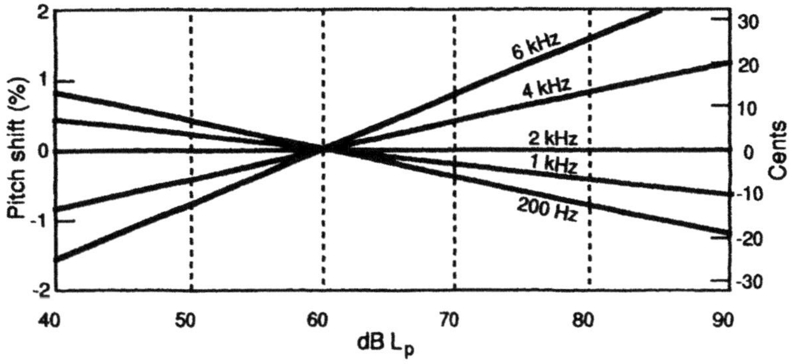

Figure 140. Pitch and Level Relationships, I

141. Pitch and Level Relationships, I

The data presented in Figure 141 were developed by Stevens, and it is clear that they show pitch shifts about twice those determined by Terhardt. The results are for a single subject who was "a good responder."

Today, the data of Terhardt are generally used. Note however that both Stevens and Terhardt identify 2 kHz as the approximate dividing point below which pitch decreases with level and above which pitch increases with level. This observation has been referred to as "Stevens's law."

Reference: Stevens (1935)

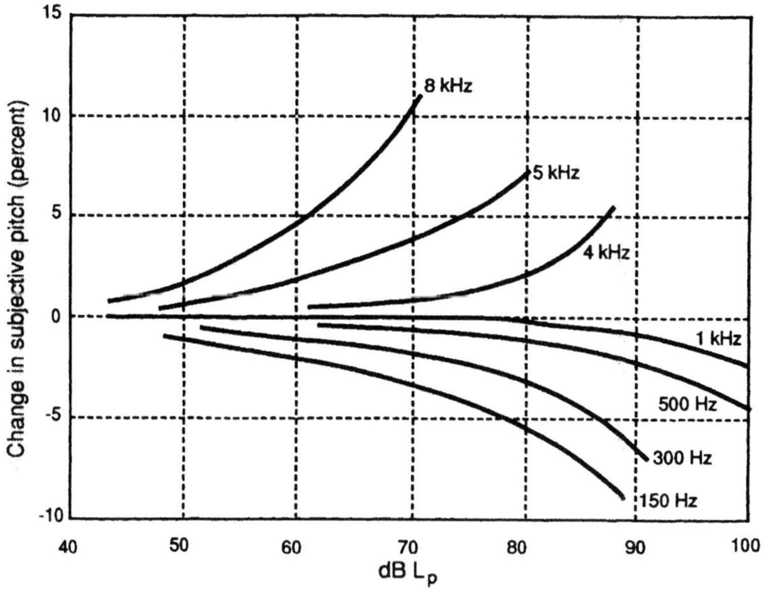

Figure 141. Pitch and Level Relationships, II

142. FREQUENCY AND PITCH RELATIONSHIPS

The relation of pitch as a function of frequency for pure tones is shown in Figure 142. The reference point here is 1000 Hz (at a level of about 60 dB L_p) where pitch and frequency are taken as equal.

Note that a frequency of about 3000 Hz will have twice the subjective pitch of 1000 Hz. Note further that the pitch scale does not exceed about 3500 mels. Even with complex musical tones, it is difficult to assign pitch to frequencies above about 4 kHz.

In the region below 1000 Hz, a pure tone of 400 Hz will have about half the pitch of the 1000-Hz reference tone.

Reference: Stevens and Volkmann (1940)

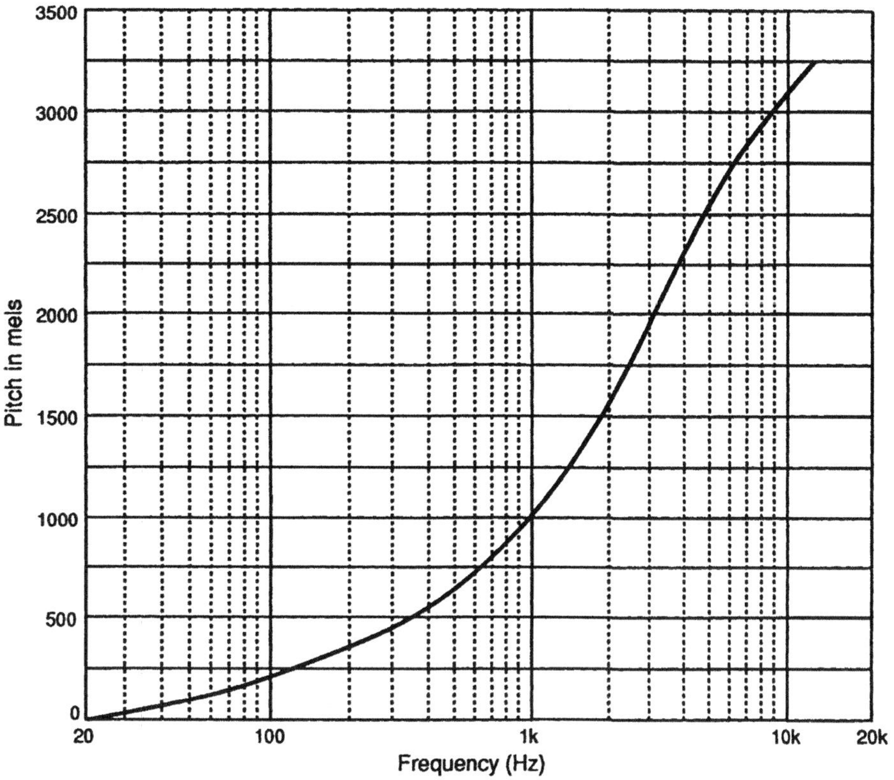

Figure 142. Frequency and Pitch Relationships

143. Critical Bandwidth

Critical bandwidth represents the maximum bandwidth, in a specific portion of the audible range, over which the ear cannot clearly separate two adjacent sine wave signals. It appears to be a property of the basic physiology of the ear and not of any further neural "postprocessing."

In Figure 143A, f_1 represents a fixed tone and f_2 a variable one. When the two are set to the same frequency, the listener will hear a single tone. Now, let f_2 change slowly in frequency, either increasing or decreasing. As this is done, the listener will hear, in essence, a single fused tone equal to $(f_1 + f_2)/2$, and there will be a beat (undulation) in the sound equal to the difference between f_1 and f_2. As long as the beats are 10 Hz or less, they will be perceived clearly; but higher beat rates will be perceived as a roughness in the sound.

When the difference between f_1 and f_2 reaches a certain value, the listener will begin to hear each tone clearly; these points along the Δf scale define critical bandwidth at the particular center frequency. Figure 143B shows the variation in critical bandwidth over the frequency range from 50 Hz to 12.5 kHz. Frequency is given along the abscissa and critical bandwidth along the ordinate.

Critical bandwidth plays an important role in our perception of music. Fundamental notions of consonance and dissonance are determined by it, and the general avoidance of close musical intervals at low frequencies is a further consequence of the observation that critical bandwidth remains fairly constant below about 200 Hz.

Reference: Roederer (1973)

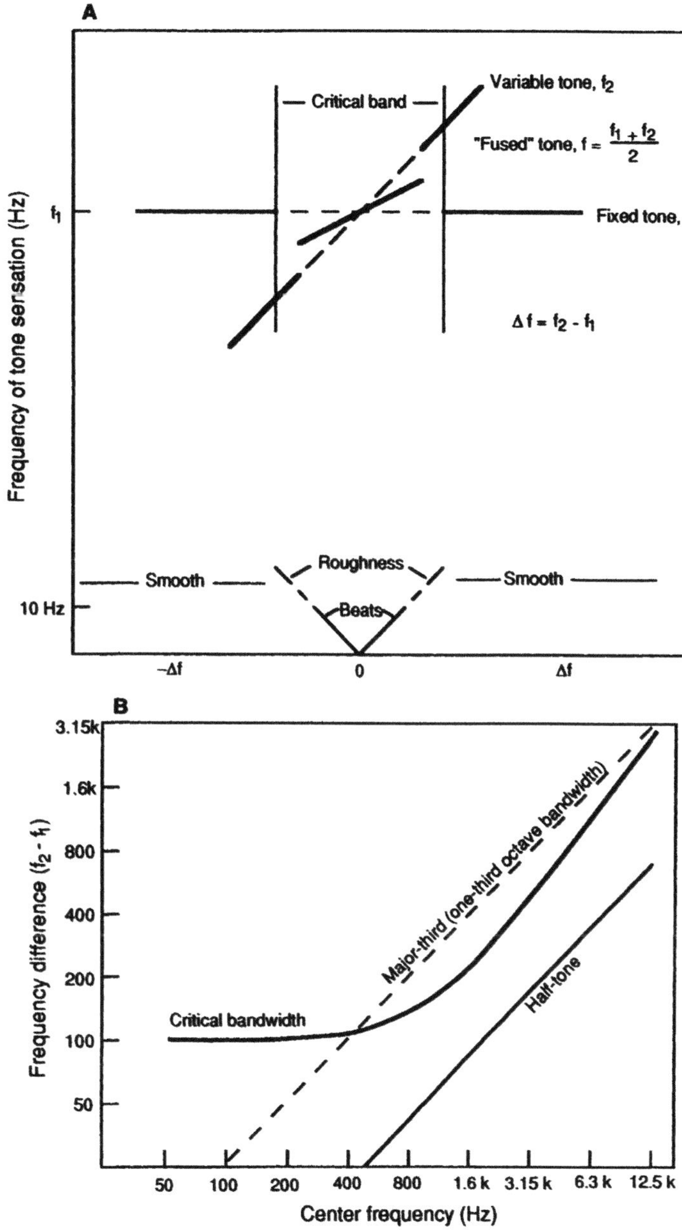

Figure 143. Critical Bandwidth

144. Annoyance Due to Echo Effects

Data generated by Bolt and Doak is shown in Figure 144. The contours indicate the thresholds of a group of subjects as they are able to perceive the presence of undesirable echoes accompanying a direct signal.

For example, 10% of the listeners reported that the presence of an echo 20 dB below the direct signal at a delay of 200 msec was annoying. Ninety percent said the same echo was annoying when it was only about 3 dB below the direct signal.

At delays shorter than about 100 msec, echoes can be louder than the direct sound and not be judged to be annoying.

Beranek refers to the Bolt and Doak data as "a tentative (and largely unsubstantiated) criterion for annoyance due to echo." Nevertheless, the data certainly define a general trend and have been of use for many years.

Before the general availability of digital signal delay devices, the 50% Bolt and Doak contour was accepted as the threshold at which either signal delay or some kind of acoustical treatment had to be considered in the successful implementation of speech reinforcement systems. Today, the ready availability of signal delay devices ensures their specification any time a delay or "time alignment" problem is detected.

References: Beranek (1954)
Bolt and Doak (1950)

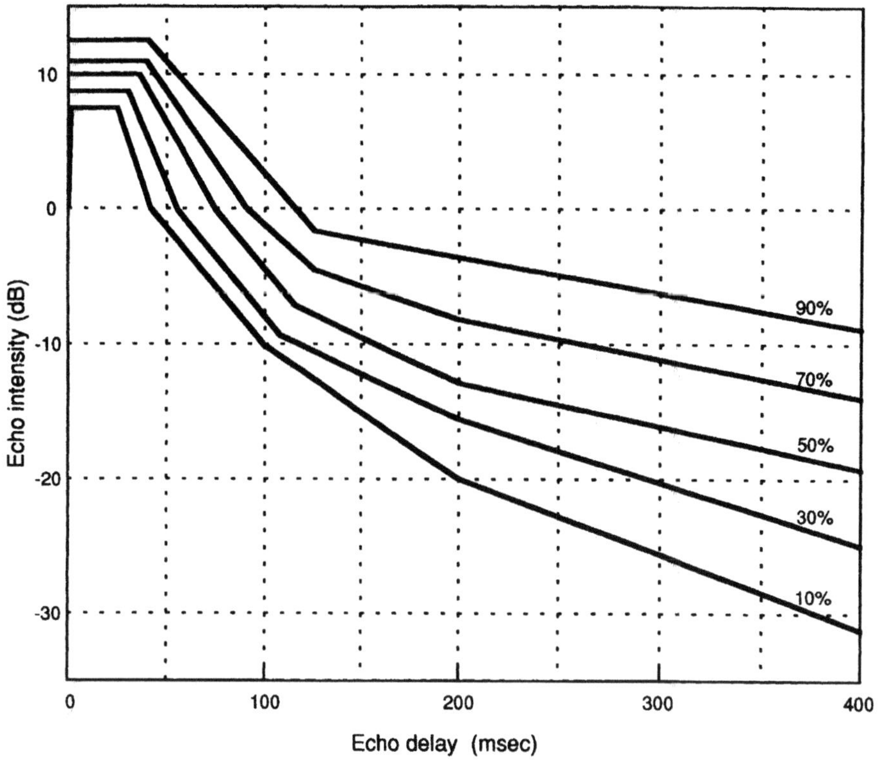

Figure 144. Annoyance Due to Echo Effects

145. Blauert and Laws Criterion for the Audibility of Signal Group Delay

Blauert and Laws have determined the basic audibility limits of delay effects in a wide variety of music and speech signals. The audible thresholds were determined with variable signal delays based on octave centers from 500 Hz to 8 kHz. It is important to state that the delays were presented to the listeners with no interaction whatever with the signal spectrum. The data should not be extrapolated beyond the frequencies presented in Figure 145.

The data are widely used and widely misunderstood. Many loudspeaker designers state that the slightest fore-aft misalignment of, say, a midrange transducer in a multiway system is audible to any listener. This may, in fact, be the case; but the culprit is not group delay as such, but rather substantial changes in signal spectrum in the overlap regions between the midrange device and its adjacent transducers due to interference effects.

What the Blauert and Laws data say in effect is that it is not necessary to design loudspeaker systems that have zero group delay over their bandpass, this despite the fact that many designers do it routinely.

Reference: Blauert and Laws (1978)

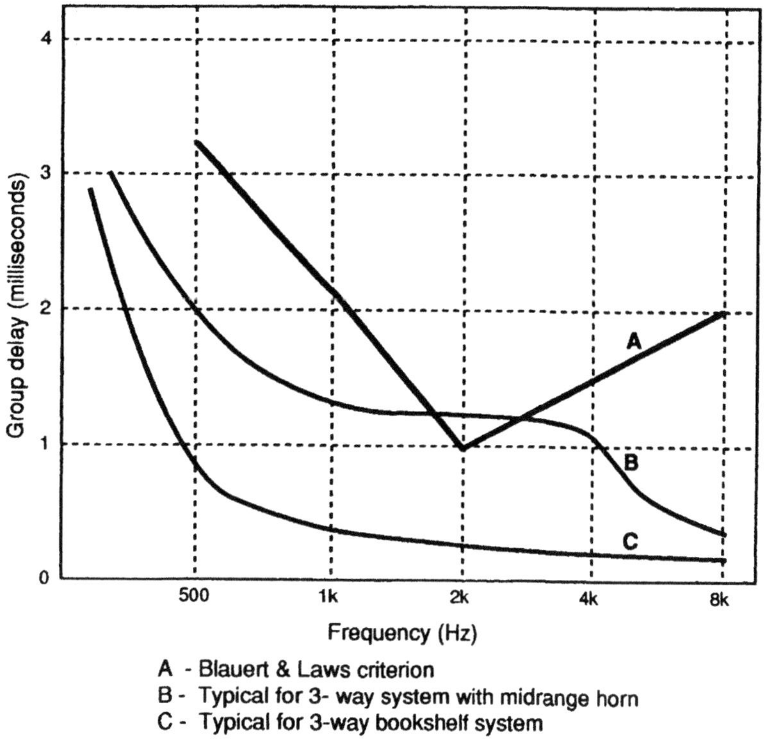

Figure 145. Blauert and Laws Criterion for the Audibility of Signal Group Delay

146. Optimum Reverberation Time as a Function of Room Volume and Usage

The data presented in Figure 146 show trends in reverberation time as it is generally influenced by room volume and usage. Since this determination is broadly subjective, we should not be surprised to see a good deal of variability in it.

The data shown in Figure 146A were developed by Knudsen and Harris, and that shown at Figure 146B by Backus. The general trend with each set of data is that, as room volume increases, the desired reverberation increases as well. The reason for this is that the direct-to-reverberant ratio is as important as the reverberation time itself; and in larger spaces, the reverberation time must increase if the direct-to-reverberant ratio is to remain fairly constant.

For this reason, it is perhaps best to justify the recommendations made by Backus for romantic, classical, and modern music as upper limits on reverberation time, regardless of the direct-to-reverberant ratio.

References: Backus (1969)
Beranek (1954)
Knudsen and Harris (1950)

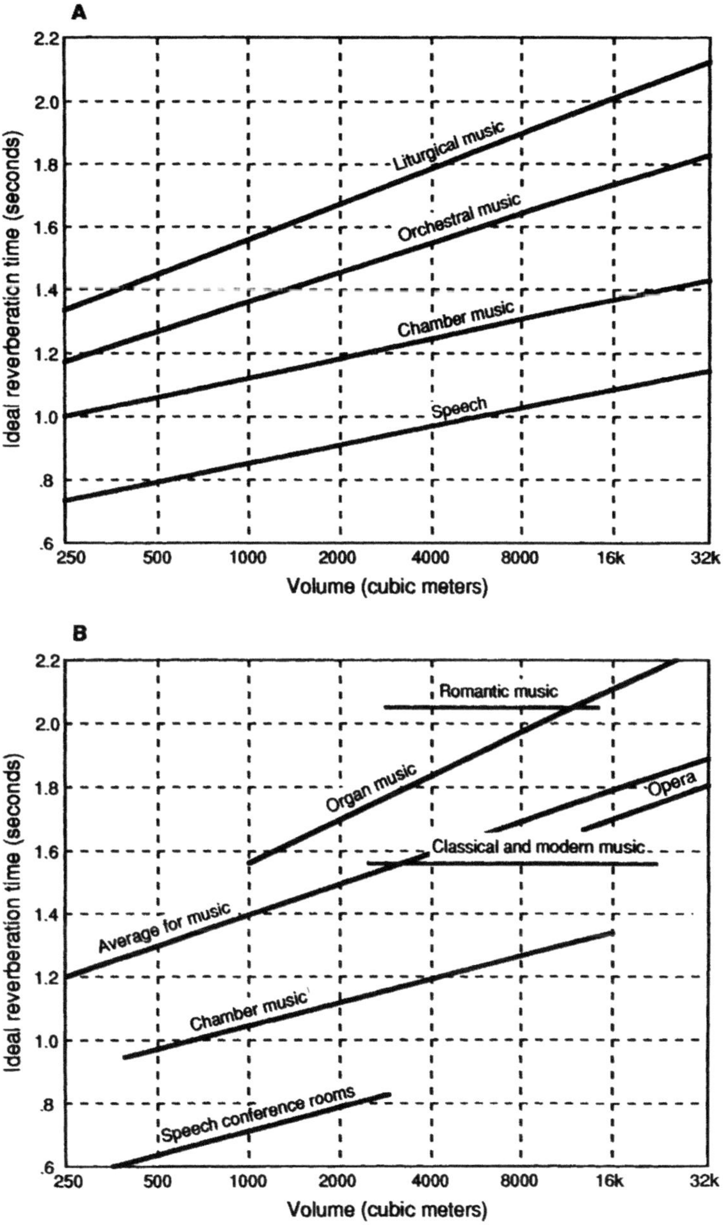

Figure 146. Optimum Reverberation Time as a Function of Room Volume and Usage

147. Optimum Reverberation Time as a Function of Frequency

The data presented in section 146 may be used to target a specific reverberation time depending on room size and use; that value is nominally the reverberation time at 500 Hz. For music, it is desirable for the reverberation time to be longer at lower frequencies by a factor of about 1.5.

The general recommendation is as shown in Figure 147A. However, in large spaces, a reduction of reverberation time above about 3 kHz may be inevitable, due to the excess absorption of air. In no case should the effect be greater than that shown in Figure 147B.

For speech applications, such as lecture rooms and the like, many acousticians feel that there should be no increase in low-frequency reverberation time.

The recommendations shown in Figure 147C are suggested for motion picture applications.

References: Beranek (1954)
Dolby Laboratories (1993)
Knudsen and Harris (1950)

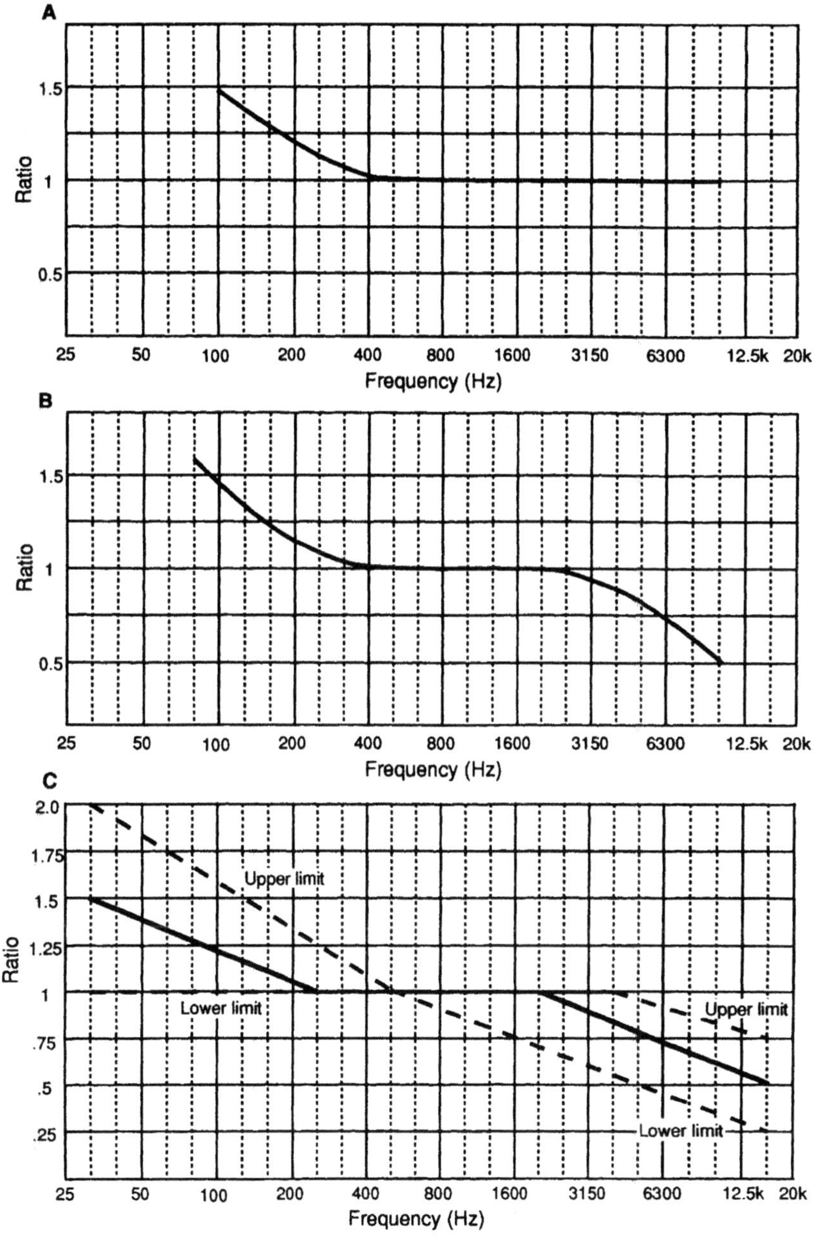

Figure 147. Optimum Reverberation Time as a Function of Frequency

148. Subjective Effects of First Reflections in a Concert Hall

The data presented in Figure 148 were developed by Barron. They show the subjective effect of first reflections in a concert hall arriving at the listener at an angle of 40° relative to direct sound arriving from the front of the listener.

Reflections are most beneficial in conveying a sense of spaciousness when they are in the range of 40 msec or longer and at appropriate levels relative to the direct sound. At shorter delays, there may be tone coloration; this is due to reinforcements and cancellations in the spectrum.

Very early reflections may cause image shift if they are high enough in level; late reflections may cause distinct echoes if they are too high in level.

Reference: Barron (1971)

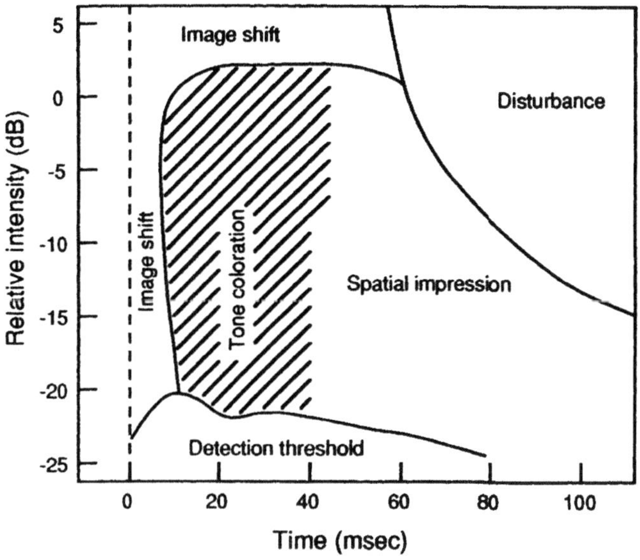

Figure 148. Subjective Effects of First Reflections in a Concert Hall

149. Binaural Lateral Masking

In Figure 149A, S_1 is a fixed source of sound and S_2 is a movable one that can be adjusted at any angle ϕ around the listener. Assume that S_1 and S_2 are radiating mutually incoherent signals, with S_1 being the primary signal.

If ϕ is made equal to zero, both S_1 and S_2 will be radiating from the same position, and we can establish as a zero dB reference that level of S_2 that will be just audible to the listener.

Then, S_2 is moved around the listener; at each position, a new adjustment of the S_2 level is made so that it is just audible to the listener.

These levels are shown in the bold curve in the Figure 149B. They indicate that the listener is far more sensitive to the level of S_2 by about 23 dB when the position of S_2 is about 90° relative to the forward position. In other words, relatively soft sounds arriving from the sides of a listener in the presence of a louder sound directly ahead will be much more apparent than those same sounds would be if they were radiated from the same forward direction as the louder sound.

The implications for stereophonic recording are very important; if two sounds are panned to the same position on the stereo stage, their levels will have to be more or less equal if they are to be heard equally. If the signals are widely separated—say, into both left and right channels—then the mixing engineer will have far greater leeway in adjusting their levels, while allowing each to be heard clearly.

References: Blauert (1983)
Damaske (1967)

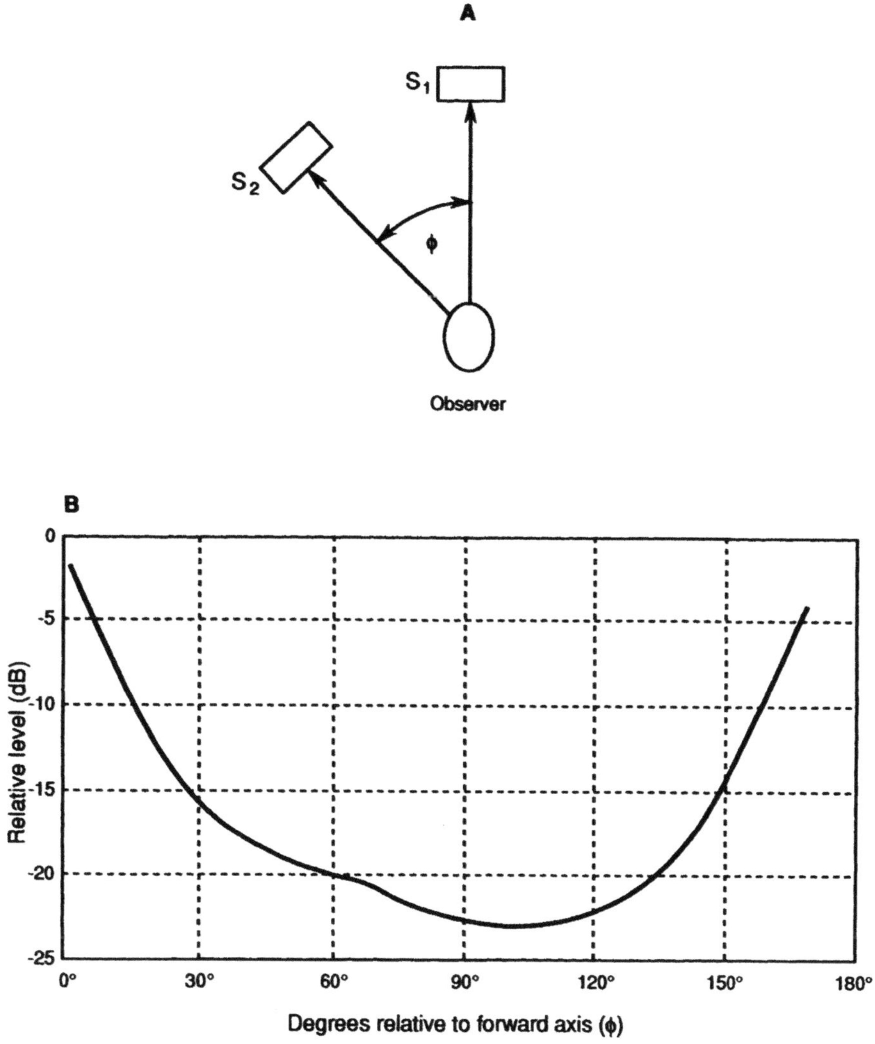

Figure 149. Binaural Lateral Masking

150. Stereophonic Localization: Franssen's Data

The data in Figure 150 demonstrate some aspects of the tradeoffs between delay and amplitude cues in determining stereophonic localization.

The point at the center of the graph indicates equal amplitudes at the two channel outputs, with no time offset between them. For a single microphone panned equally into both channels, this will correspond to a center phantom image.

If both channels are at the same level and the left channel leads the right by, say, 4 msec, then localization will clearly be at the left loudspeaker. If under this condition the level of the left channel is dropped by 9 dB, then localization will again be at the center of the stereophonic stage.

The line marked A–A′ represents the operating range for a spaced stereo microphone array. The line marked B–B′ represents the operating range for a coincident stereo microphone pair. The data apply to broadband signals in a normal program context.

According to Blauert, "the curves have not been sufficiently verified by measurement to be useful other than for qualitative study."

References: Blauert (1983)
Franssen (1963)

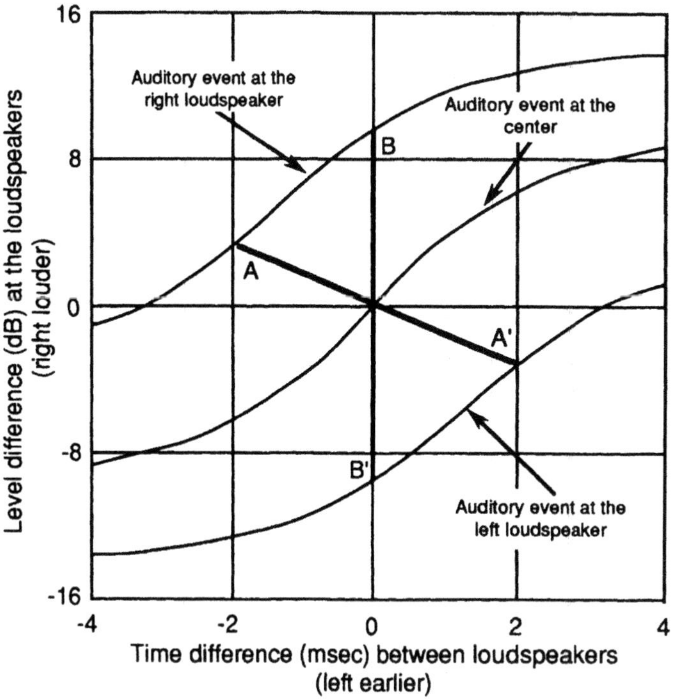

Figure 150. Stereophonic Localization: Franssen's Data

151. The Precedence Effect (Haas Effect)

In the schematic shown in Figure 151A, if the adjustable delay is set at zero and equal signal levels fed to the loudspeakers, a listener will localize the event at the center of the loudspeaker array. As delay is introduced, localization quickly moves to the earlier of the two loudspeakers.

Delays up to about 5 msec can be compensated for by decreasing the level of the earlier loudspeaker by about 10 dB. As shown in Figure 151B, the delay-amplitude tradeoff is approximately linear over this range. For delays in excess of 5 msec, a constant level difference of about 10 dB will maintain the image at the center position.

As delays increase to 25 msec and above, the listener will begin to hear distinct signals at each loudspeaker.

Note that there is good correspondence between the data shown here and Franssen's data in section 150.

References: Blauert (1983)
Haas (1972)

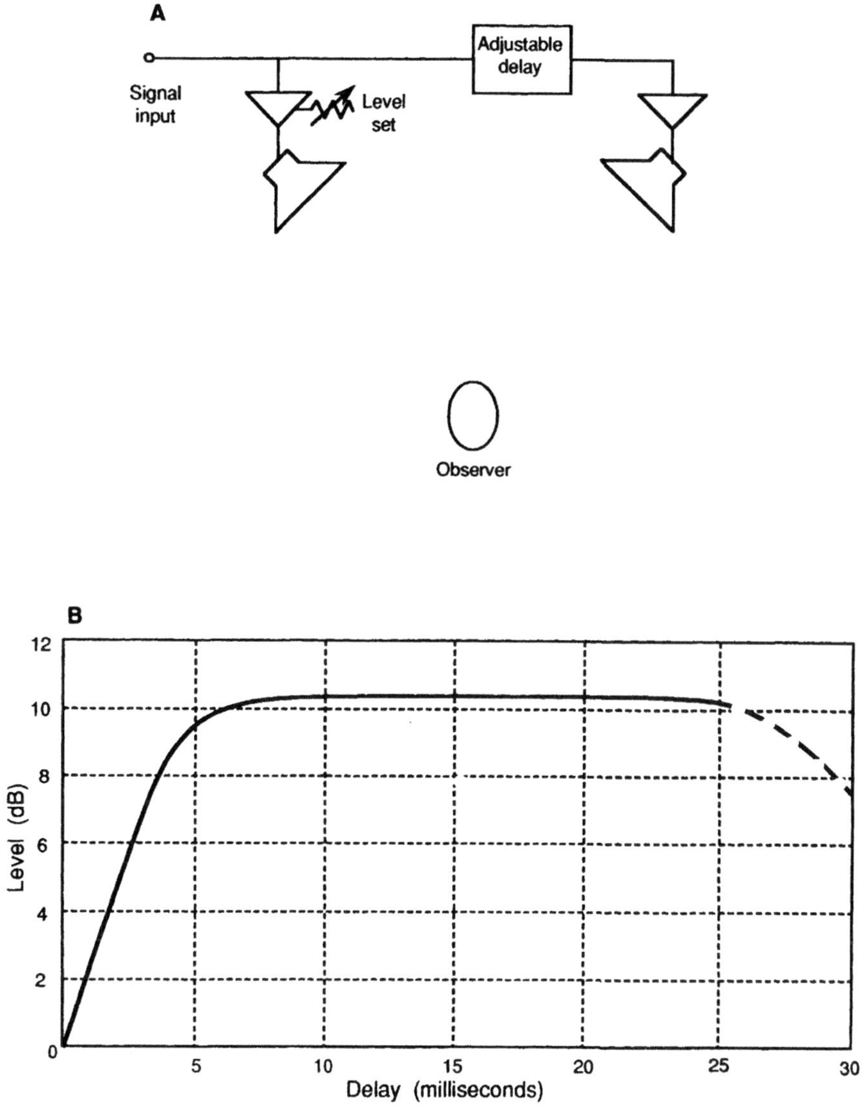

Figure 151. The Precedence Effect (Haas Effect)

152. BAUER'S STEREOPHONIC LAW OF SINES

In coincident stereophonic microphone systems, the approximate location of an event on the stereo stage can be determined by the following equation:

$$\sin \theta_1/\sin \theta_A = (S_L - S_R)/S_L + S_R) \quad \ldots\ldots\ldots\ldots \quad 152.1$$

where θ_1 is the angular position of the image from the center line and $2\theta_A$ is the included angle of the stereophonic loudspeaker array as seen by the listener. This relationship is illustrated in Figure 152.

S_L and S_R are, respectively, the sound pressures at the listener produced by the left and right loudspeakers.

References: Bauer (1956)
Clark, Dutton, and Vanderlyn (1958)
Eargle (1992)

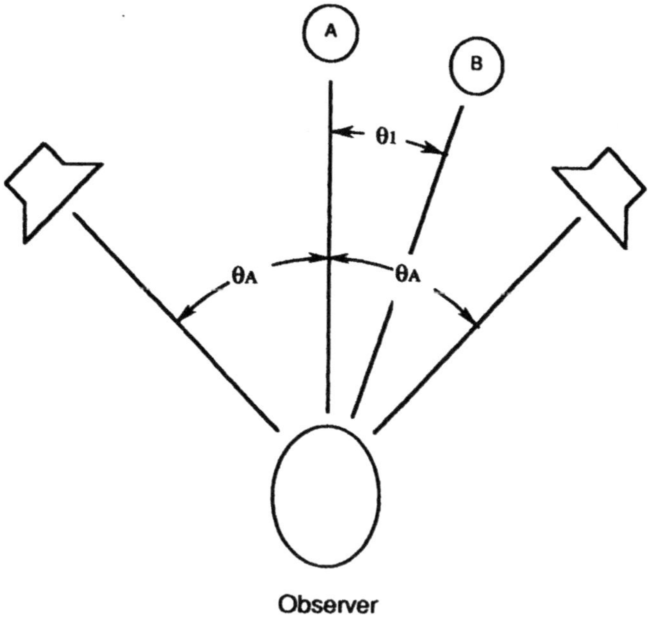

Figure 152. Bauer's Stereophonic Law of Sines

153. Pressures and Pressure Levels Generated by a Variety of Sound Sources

Figure 153 gives sound pressures and their corresponding levels for sounds and acoustical conditions spanning a range of more than 200 dB.

The significance of 194 dB L_p as equivalent atmospheric pressure is that the negative going half of the waveform will cancel the static positive atmospheric pressure and a shock wave will develop. Long before that level has been reached, however, there will be much distortion of acoustical signals due to thermodynamic nonlinearities.

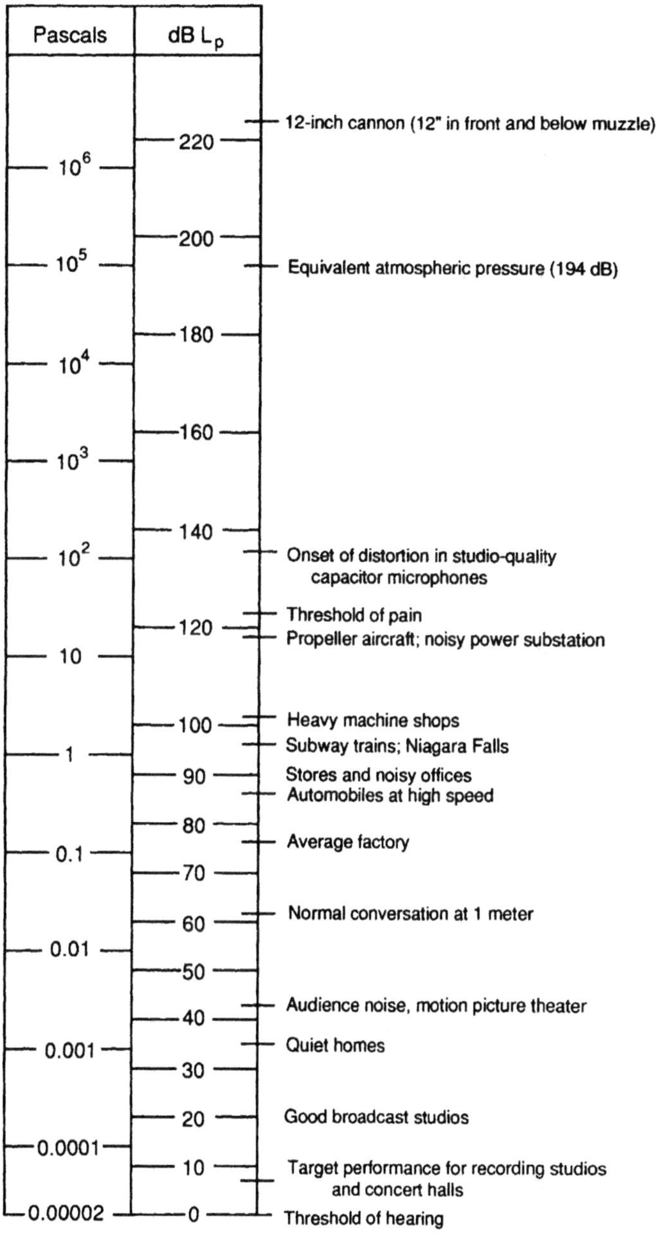

Figure 153. Pressures and Pressure Levels Generated by a Variety of Sound Sources

154. Typical Male Speech Spectra

Figure 154 shows octave band pressure levels of male speech measured at a distance of 1 meter. Note that the total range of speech levels is quite wide, spanning a range of about 55 dB at 1 kHz.

Above normal speaking level, the spectrum rises most prominently at 1 kHz and much less at other octave band centers. The level in the 125-Hz band appears to remain fairly constant at higher levels, giving elevated speech a characteristically bright sound.

The curves presented in Figure 154 represent smoothed average values for a typical male subject. The descriptive terms used in the graph have no precise meaning; levels that are "normal" for one talker may not be the same for another.

The A-weighted values associated with the spectra presented here are:

Vocal effort	*Level (dB(A))*
Maximum	88
Shout	82
Very loud	74
Raised	65
Normal	57
Relaxed	50
Whisper	40

References: Harris (1979)
Western Electro-Acoustic Laboratory (1959)

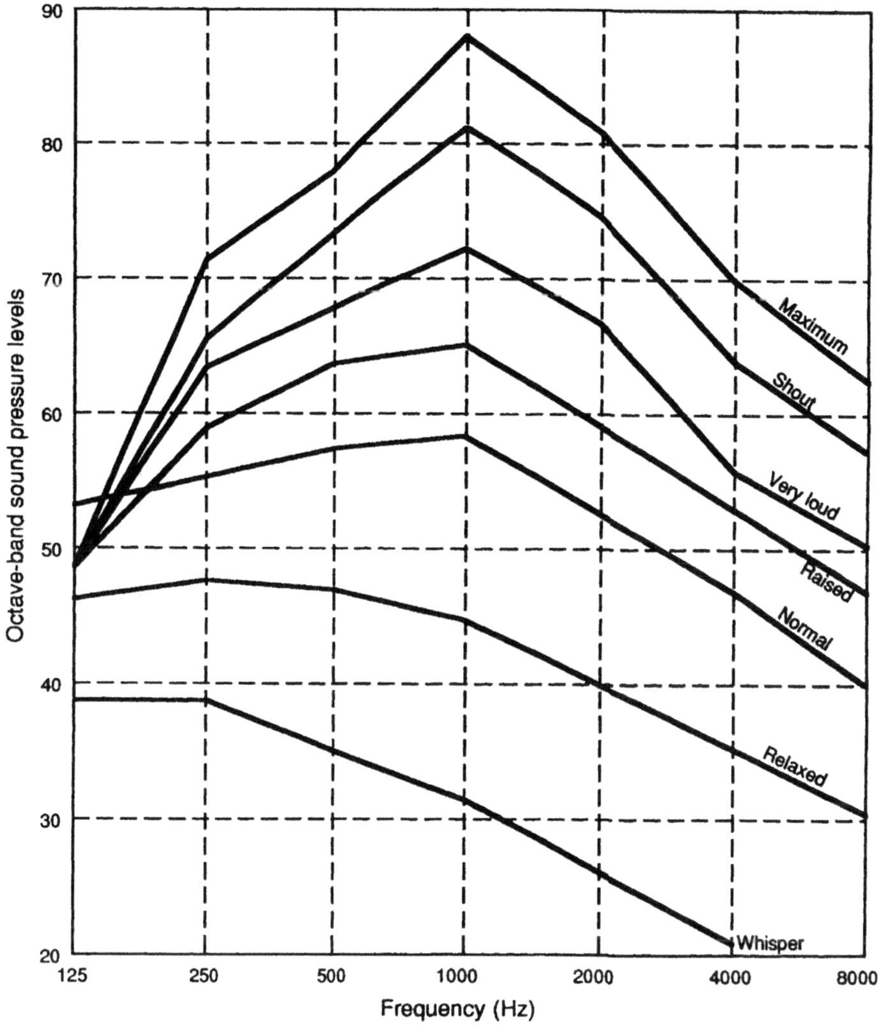

Figure 154. Typical Male Speech Spectra

155. Hearing Threshold Shift as a Function of Age

For persons with normal hearing who are not subjected to unusually high noise levels for any substantial length of time, there is a progressive shift of hearing threshold with age. The loss is known as "presbycusis." Women generally fare better than men, although there is a good bit of variation in the threshold shift in the population at large.

The curves shown in Figure 155 are often misinterpreted as indicating that hearing loss is a kind of "high frequency rolloff," such as we might perform via a tone control on a modern sound-reproducing system. This is not the case; once a signal has exceeded the hearing threshold, it will be as loud to an older person as to a young one.

The onset of threshold shift can begin as early as the mid-20s, but there are much older individuals whose hearing seems to defy the odds.

The modern age subjects many people to a wide variety of noise in the workplace, as well as in the course of simply getting to and from the workplace. The great popularity of rock concerts carries with it obvious risks; part of the blame for threshold shift can be placed squarely with those who attend these events.

References: Harris (1979)
Spoor (1967)

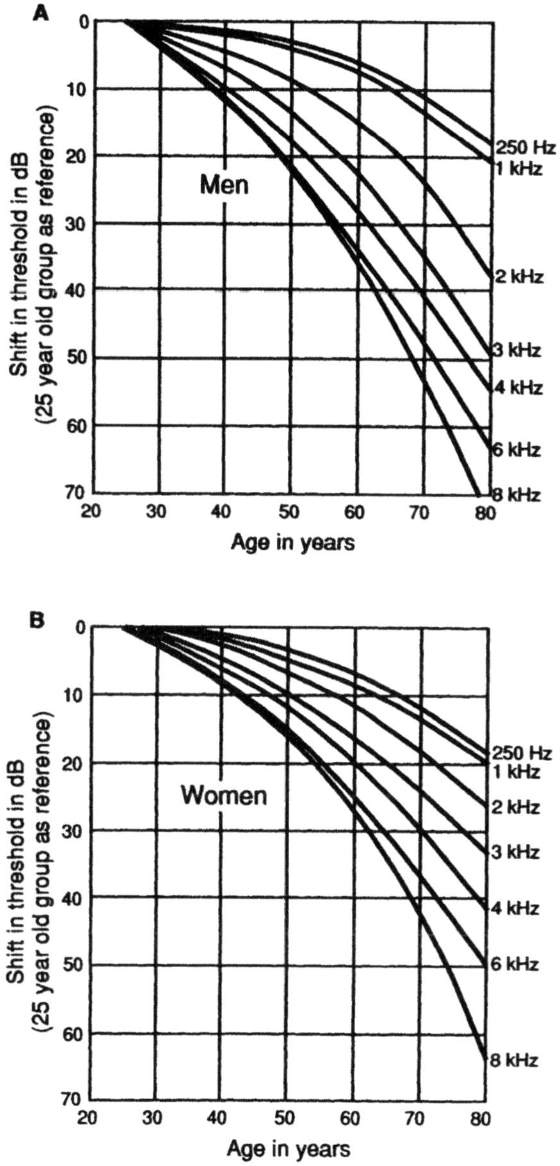

Figure 155. Hearing Threshold Shift as a Function of Age

PART SIX

MUSICAL INSTRUMENTS

156. Frequency Ranges of Musical Instruments and the Human Voice

Figure 156 shows the ranges of fundamental frequencies and harmonics associated with various orchestral instruments. The normal fundamental ranges of male and female singing voices and the piano are also given.

The normal piano keyboard is given for reference, as are frequencies based on the normal tuning of A-440 Hz.

Standard nomenclature of the notes of the scale is used; C0 represents "low C" with a fundamental of 16.3 Hz and C4 represents "middle C" with a fundamental of 261.63 Hz. This nomenclature is fairly common in musical acoustics, but other nomenclature is used by organ builders.

The crosshatched low-frequency extension of the piano keyboard represents the extension down to low C found only on the Bösendorfer Imperial Grand piano. The treble extension up to E8 has never been observed by the author.

Note overall that most musical fundamentals are to be found well below 4 kHz, with harmonics covering the upper two octaves out to 16 kHz. Although the normal upper-frequency limit of hearing (and, by extension, music) is often stated to be 20 kHz, there is no assurance that this is necessarily the case. Many musical instruments routinely produce frequencies well beyond 20 kHz.

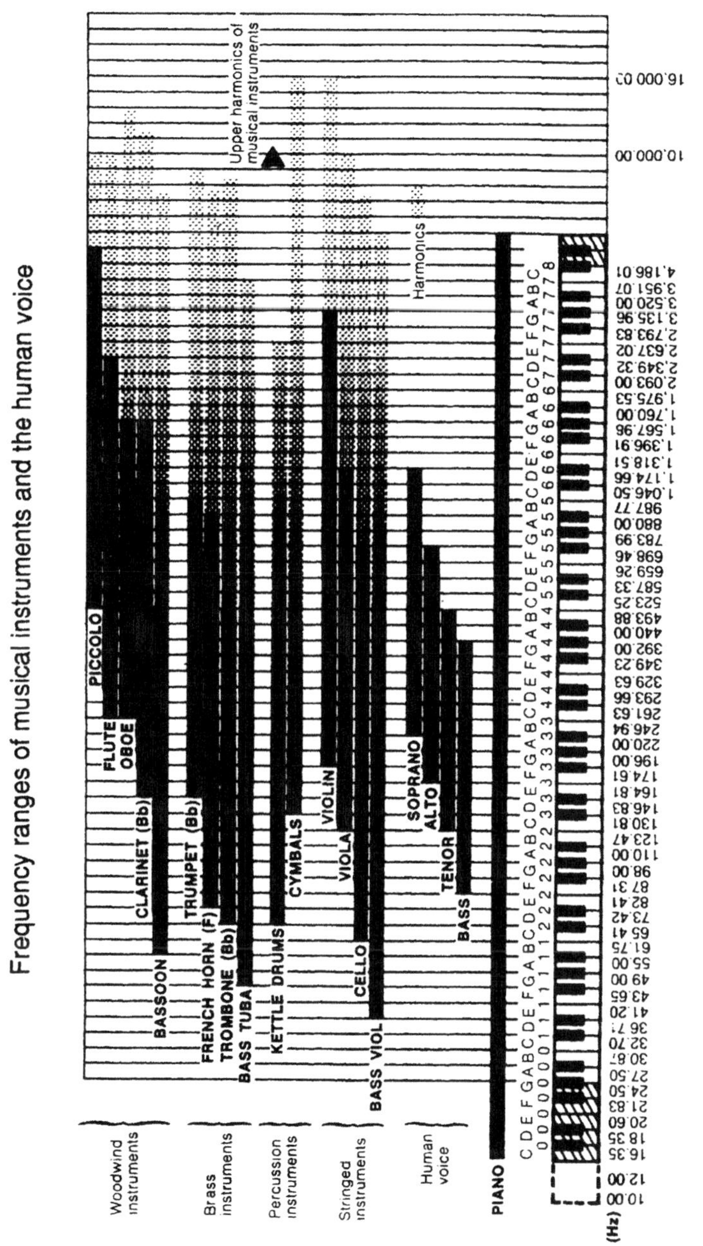

Figure 156. Frequency Ranges of Musical Instruments and the Human Voice

157. Dynamic Ranges of Wind and String Instruments

The ranges shown in Figure 157 represent the span from the lowest possible levels that can be continuously produced by these instruments up to the highest levels that can be produced. In most cases, the lowest and highest output capabilities will not be in the same frequency range. In general, wind instruments exhibit their highest output capability at higher frequencies.

Percussion instruments have not been included here, inasmuch as there are no fundamental limits at lower levels, at which point the sound is simply damping itself slowly to zero output.

A-weighting has been used for the lower levels shown, and all levels are referred to a distance of 1 meter.

Note, in general, that the dynamic ranges of most of these instruments are about 30–40 dB. Notable exceptions are the clarinet (about 50 dB) and the French horn (about 65 dB).

References: Eargle (1990)
Meyer (1978)

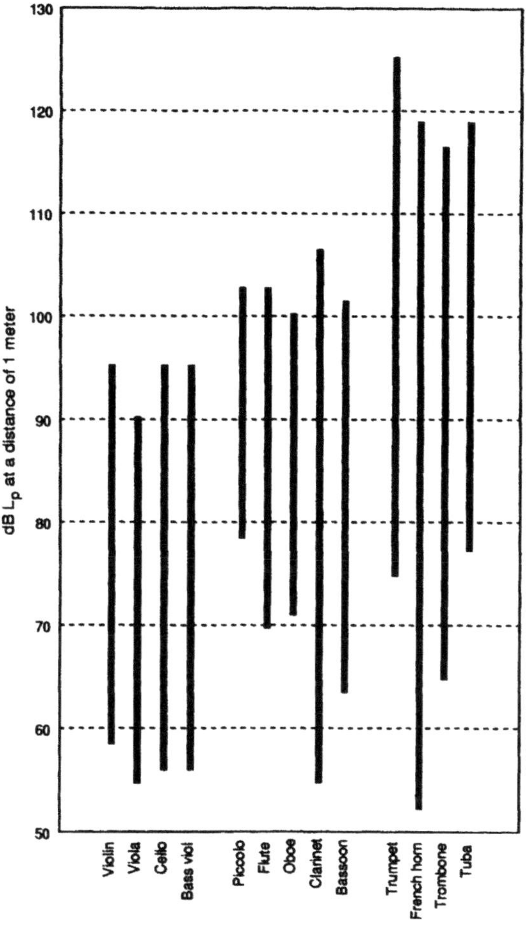

Figure 157. Dynamic Ranges of Wind and String Instruments

158. Directional Properties of Brass Instruments

Directional radiation from the bells of brass instruments is fairly simple. The data presented in Figure 158 are adapted from the work of Benade. The four polar plots shown here are based on four octave-related frequencies. The reference values of f are as follows:

Instrument	*Value of f*
Trumpet	500 Hz
Trombone	250 Hz
Tuba	167 Hz

The French horn is difficult to fit into this scheme, inasmuch as the player's right hand is always inserted into the bell for added control of timber and pitch.

By referring to section 156, you can see that the highest fundamental frequency normally produced by the trumpet is just below 1 kHz; this frequency corresponds to 2f in the polar diagram. Thus, it is easy to see that all trumpet fundamentals are relatively nondirectional; it is the higher harmonics that give the instrument its pronounced directionality at higher playing levels.

References: Benade (1976)
Eargle (1990)
Meyer (1978)

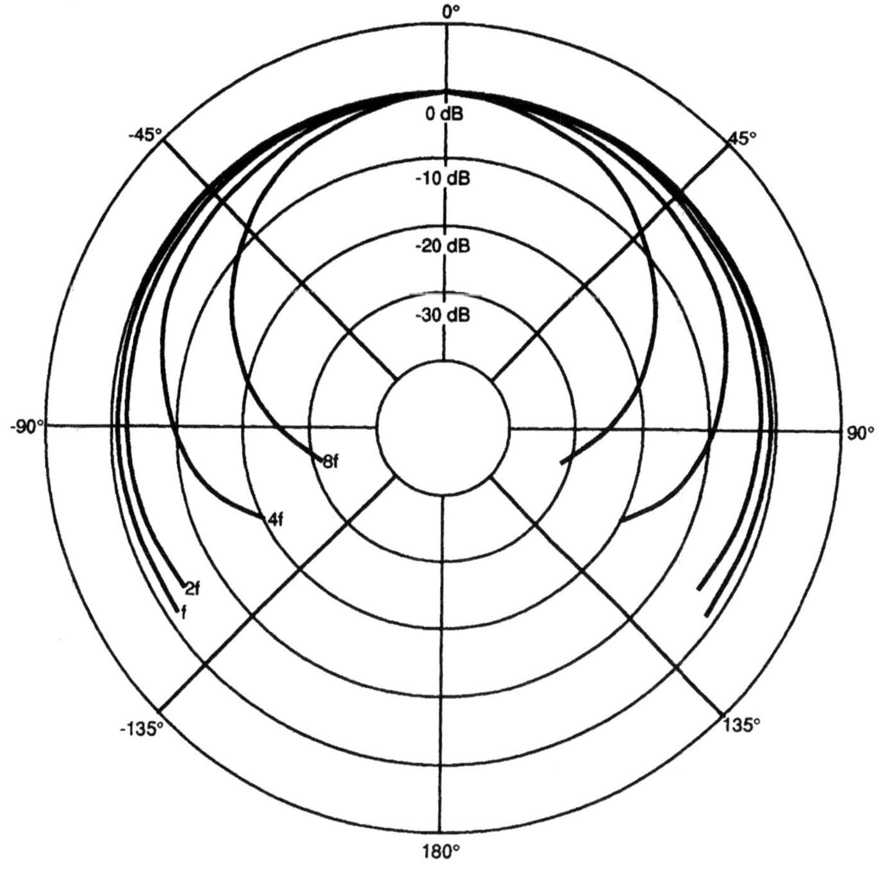

Figure 158. Directional Properties of Brass Instruments

159. Directional Properties of Woodwind Instruments

The directional characteristics of woodwind instruments are much more complicated than those of brass instruments because the various open tone holes, as well as the bell, influence the behavior. Benade has shown that the directional characteristics, below a certain cutoff frequency, become essentially omnidirectional. At the cutoff frequency the radiation is primarily to the sides of the instrument, with very little along the axis of the bell. At frequencies progressively higher than the cutoff frequency, the radiation pattern gradually becomes more pronounced from the end of the instrument along the axis of the bell. Benade states that the cutoff frequency is a function of the bore diameter, tone hole spacing, and tone hole depth.

The basic characteristic is there for all woodwind instruments; however, instruments such as the saxophone and bassoon, which double back on themselves to a certain extent, confuse the analysis somewhat.

The clarinet and oboe have similar cutoff frequencies of about 1500 Hz, whereas that of the English horn and alto clarinet is about 1000 Hz. The bass clarinet cutoff frequency is about 750 Hz; that of the bassoon is about 500 Hz.

The four curves shown in Figure 159 present the polar response at the cutoff frequency, 1.3 times the cutoff frequency, 2 times the cutoff frequency, and 4 times the cutoff frequency. 0° represents the direction along the axis of the instrument.

References: Benade (1976)
Eargle (1990)
Meyer (1978)

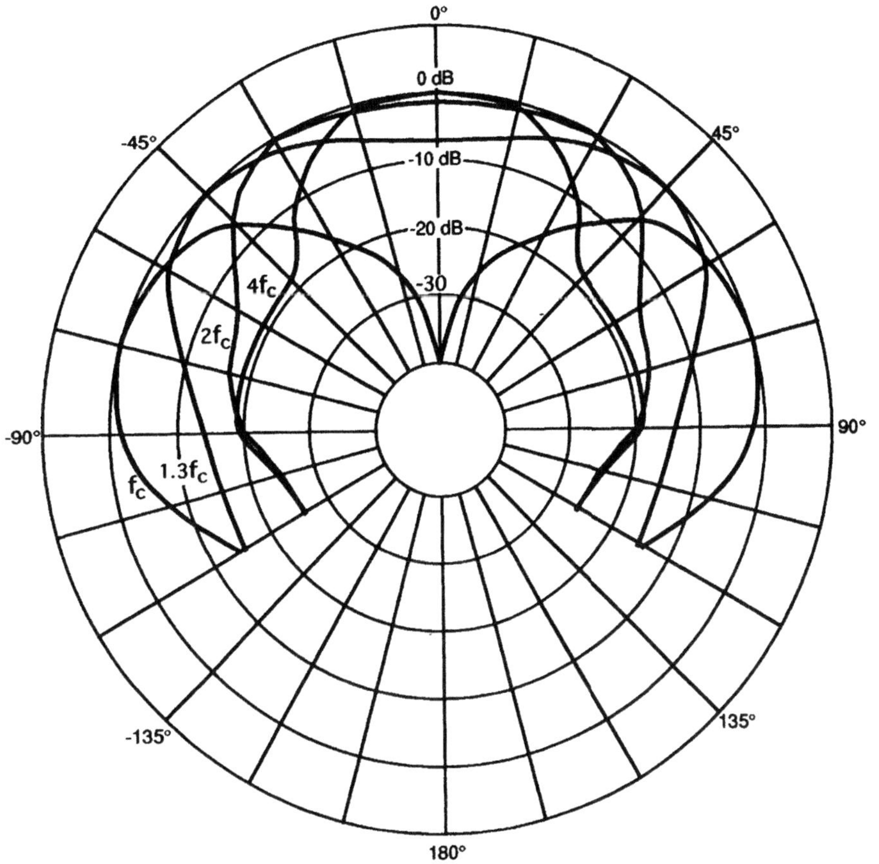

Figure 159. Directional Properties of Woodwind Instruments

160. Directional Properties of String Instruments

The directional properties of string instruments are quite complex. Figure 160 simplifies the presentation by showing the approximate radiation patterns in ocatve bands averaged over 15° angular increments. In each of the octave bands, the highest level has been normalized to 0 dB.

Sound radiation from the violin in the vertical plane is shown in Figure 160A. Note that the radiation is quite broad in the 500-Hz band. Narrowing is evident in the 1-kHz band, and there is some tendency for the radiation pattern to broaden slightly in the 2- and 4-kHz bands. The broadening is due in part to the fact that much of the radiation at those frequencies is directly from the very small bridge of the instrument.

Sound radiation from the violin in the horizontal plane is shown in Figure 160B. Radiation is most effective in the range of 0° or to the right of the player. The obstacle formed by the player's body can clearly be seen in the range between 180° and 270°.

Sound radiation from the cello is shown in Figure 160C. Most effective radiation is in the angular range of 30–90°. Again, the obstacle formed by the player's body can be seen in the range of 180–270°.

References: Eargle (1990)
Meyer (1987)

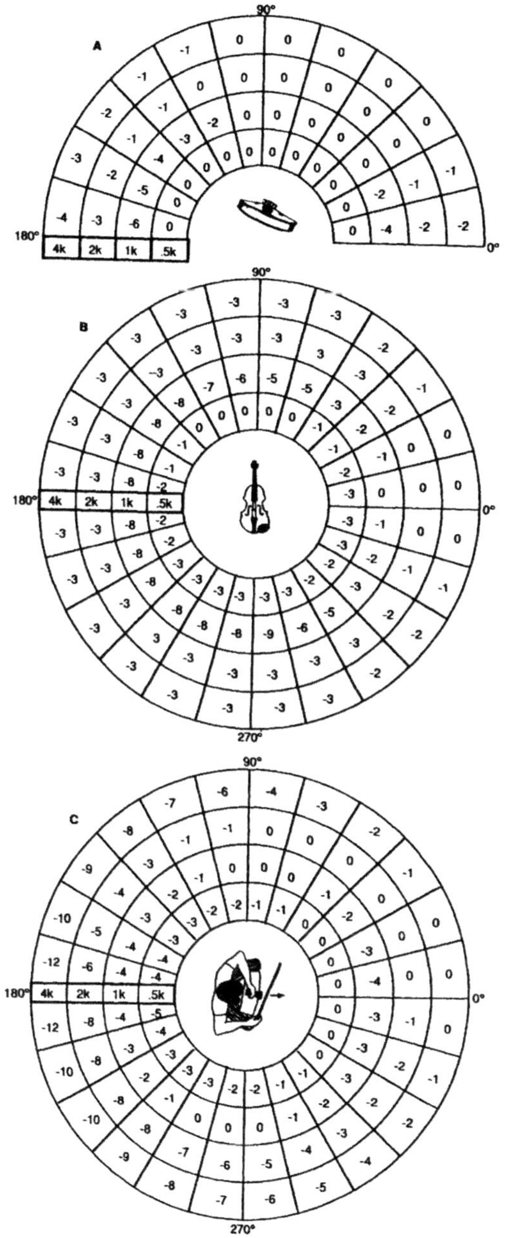

Figure 160. Directional Properties of String Instruments

161. Octave Band Spectral Amplitude Distribution, Music Sources

The data presented in Figure 161 were measured from current compact disc music sources. The curves represent the time variation of the spectra. For example, the curve labeled 1% indicates the spectrum that was exceeded only 1% of the time during the sample period. This would be indicative of the peak recorded signal. The curve labeled 50% is that spectrum that was exceeded half the time, and that marked 90% would effectively indicate the spectrum of the softest signals present in the recording.

The small dots in the upper left part of the graphs indicate the unfiltered levels in terms of time distribution. This distribution gives a good indication of the overall dynamic range of the program.

Figure 161A shows the distribution for a symphony orchestra. The rolloff above about 2 kHz is typical for musical ensembles and instruments that do not make use of amplification. Even at the highest levels, the rolloff appears to be approximately 6 dB per octave, becoming much steeper at lower levels.

Figure 161B shows the distribution for a jazz ensemble. The high-frequency portion of the spectrum is maintained flat well out to 8 kHz. This is probably a result of close microphone placement to the cymbals and wind instruments such as the trumpet and the saxophone, which maintain substantial high-frequency output at high playing levels.

Figure 161C shows the distribution for a rock group. Here, there is a preponderance of output in the 1–4 kHz range, with relatively small dynamic range.

Figure 161D shows the distribution for a piano. Note that the overall spectra tend to top out in the 250–500 Hz range, with rapid rolloff above and below that range. The running dynamic range is about 25 dB.

Figure 161E shows the distribution for a pipe organ. For this musical selection, the dynamic range is unusually small. The overall spectral shape is, however, characteristic of the instrument, which has a preponderance of low frequencies.

Reference: Greiner and Eggers (1989)

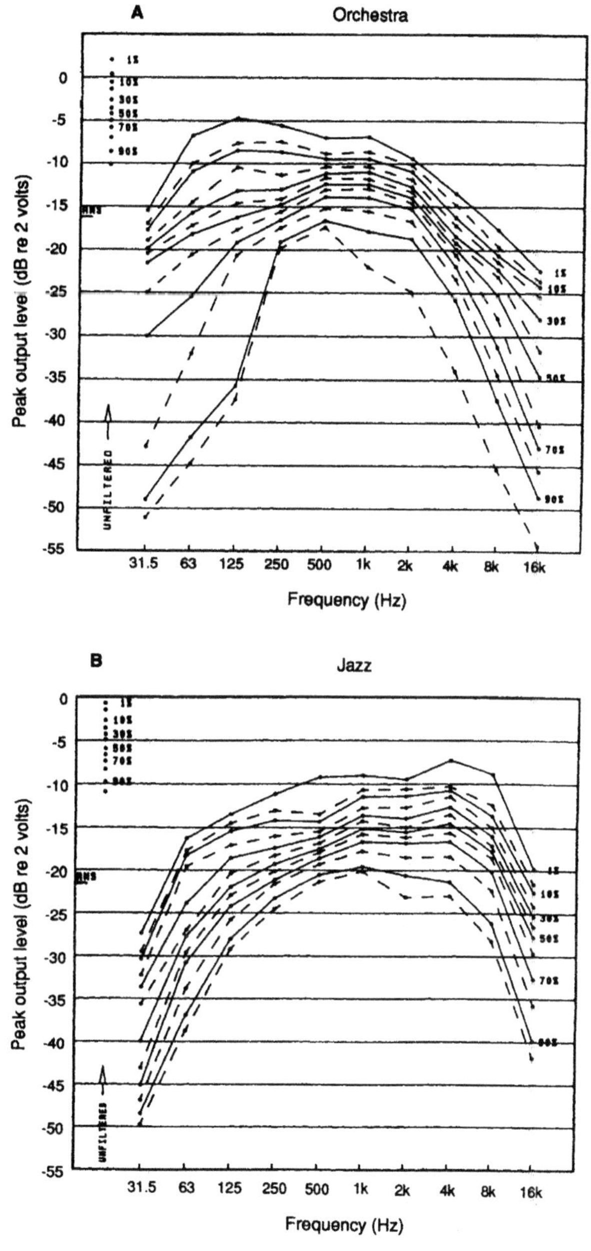

Figure 161. Octave Band Spectral Amplitude Distribution, Music Sources

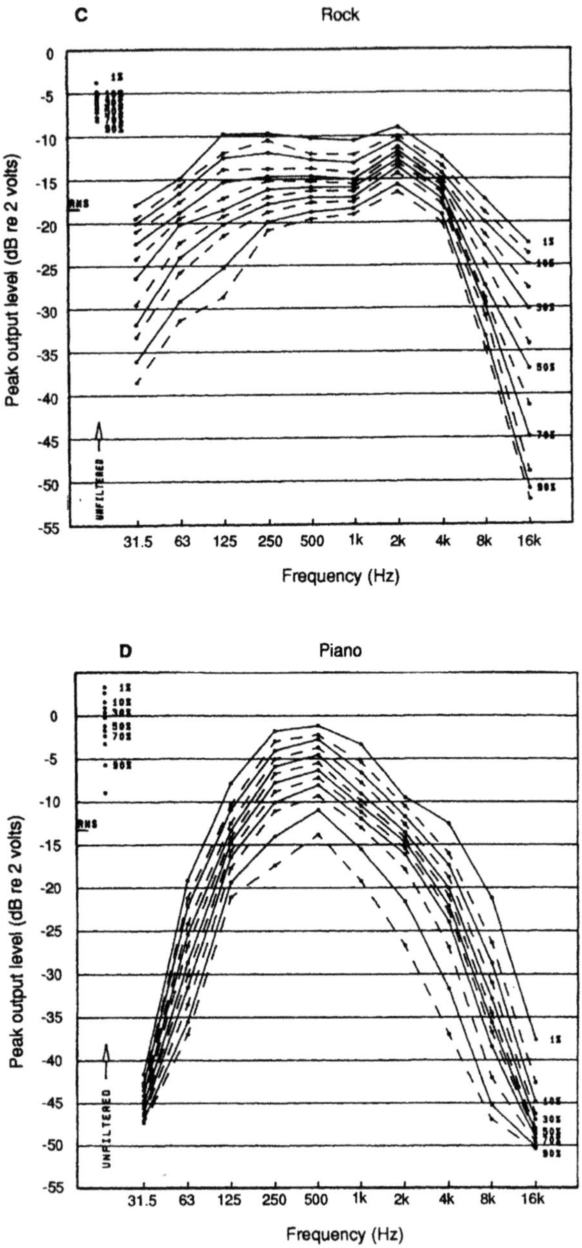

Figure 161. (Continued)

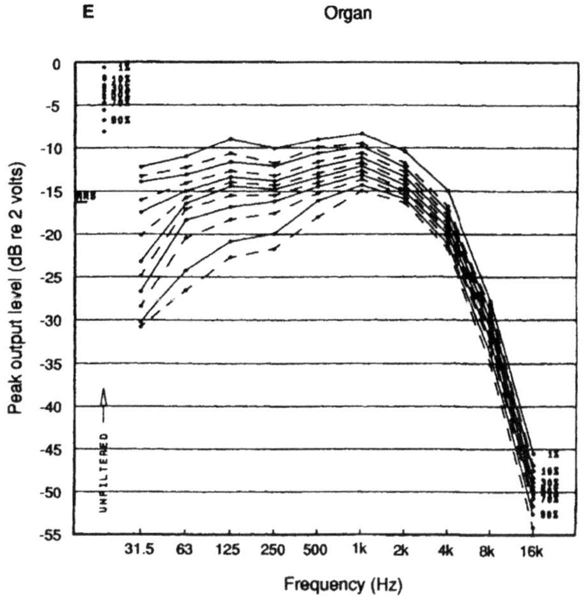

Figure 161. (Continued)

PART SEVEN

ANALOG MAGNETIC RECORDING

162. Track Width Standards for Professional Magnetic Recording

Figure 162 gives track width standards for the various formats that have been used with 6.3-mm (.25-in), 12.7-mm (.50-in), 25.4-mm (1-in), and 51-mm (2-in) tape. All dimensions are given in inches.

The relative noise levels of the various track widths are normalized here with respect to full track on 6.3-mm tape:

Format	*Relative noise level*
6.3-mm full-track	0 dB (reference)
6.3-mm two-track	–5
12.7-mm two-track	–0.7
12.7-mm three-track	–3.7
12.7-mm four-track	–4.8
25.4-mm eight-track	–4.8
51-mm 16-track	–4.8
51-mm 24-track	–7.4

References: Camras (1988)
Eargle (1992)

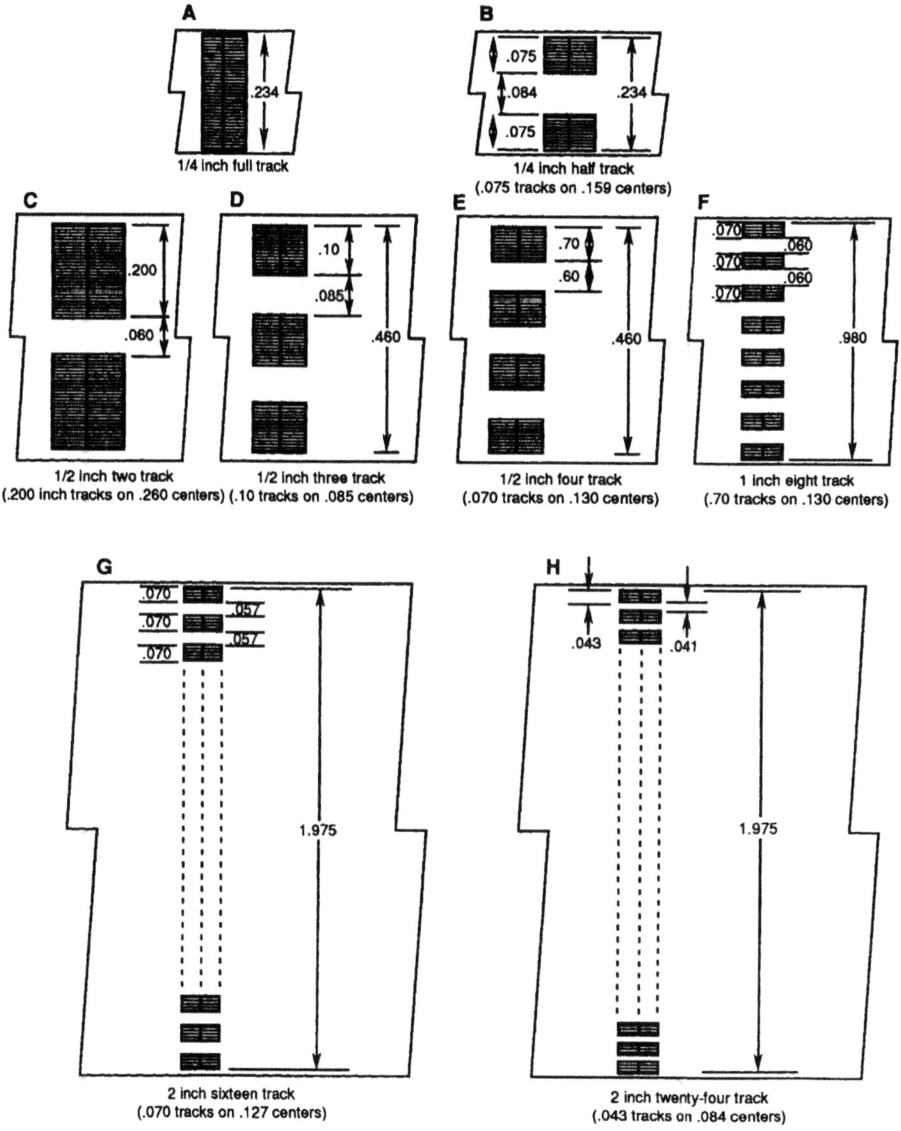

Figure 162. Track Width Standards for Professional Magnetic Recording

163. Track Widths Standards for Consumer Tape Formats

The Philips compact cassette has dominated the consumer tape recording market for at least a quarter century. The basic track format is shown at Figure 163A. Because the tape travels in both directions, there is a generous guard band between the two sets of tracks to minimize crosstalk from the adjacent program.

Earlier, reel-to-reel tape systems enjoyed some favor in the consumer marketplace. The first of these was identical to the format shown in Figure 162B. During the late 1950s, a quarter-track format was developed; it is shown in Figure 163B. Like the compact cassette, quarter-track reel-to-reel tape ran in both directions in an attempt to increase overall playing time. Unlike the cassette, the tracks were interleaved; this increased the likelihood of crosstalk.

During the brief era of quadraphonic sound, the format was adapted to recording all four channels at once, as shown at the right of Figure 163B.

From the mid-1960s to the mid-1970s, the Stereo-8 endless-loop cartridge enjoyed some success, particularly in the automobile. This was a two-track playback medium that made use of four programs distributed over eight tracks. Each pass over the loop required that the playback heads index from one program sequence to the next.

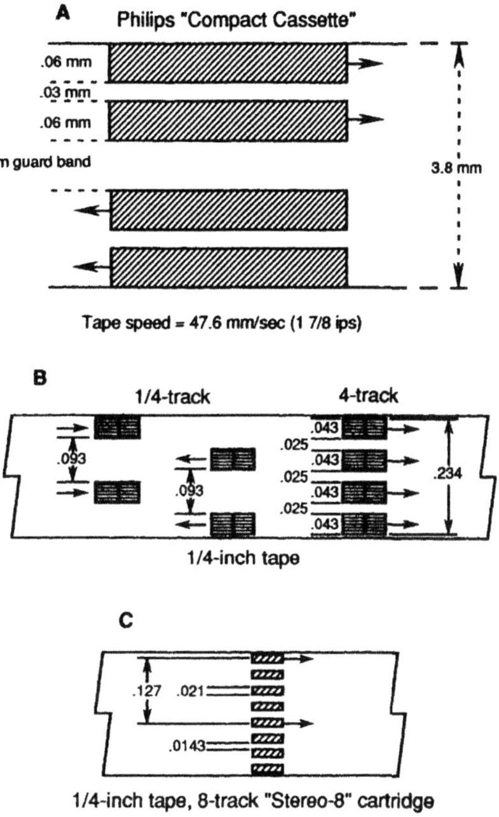

Figure 163. Track Width Standards for Consumer Tape Formats

164. Azimuth Losses in Tape Playback

The nature of azimuth loss is shown in Figure 164A. When the playback head tracks the tape at an angle different from the recording angle there will be a loss in output. Losses, shown in Figure 164B, are given by the following equation:

$$\text{Response (dB)} = 20 \log \left[\frac{\sin([\pi W \tan \theta]/\lambda)}{([\pi W \tan \theta]/\lambda)} \right] \quad (164.1)$$

where W is the track width, λ the recorded wavelength, and θ the angle of misalignment between record and playback heads in degrees. W and λ must be in the same units. (In the equation, the sine term in the numerator must be evaluated in radians.)

The nomograph in Figure 164C relates tape speed, frequency, and wavelength. A null in the playback response takes place when

$$\lambda = W \tan \theta \quad (164.2)$$

EXAMPLE: Find the azimuth loss for a track width of 2 mm with a misalignment angle of 1° for a 38 cm/sec tape at a frequency of 10 kHz.

First, determine the recorded wavelength from the nomograph at Figure 164C. By inspection, this value is 0.038 mm. Next, determine the value $W/\lambda = 2/0.038 = 52.6$.

Finally, locate the value of 52.6 along the abscissa of the graph and move upward to the 1° curve. The loss is approximately 20 dB.

References: Busby (1988)
Camras (1988)
Eargle (1992)

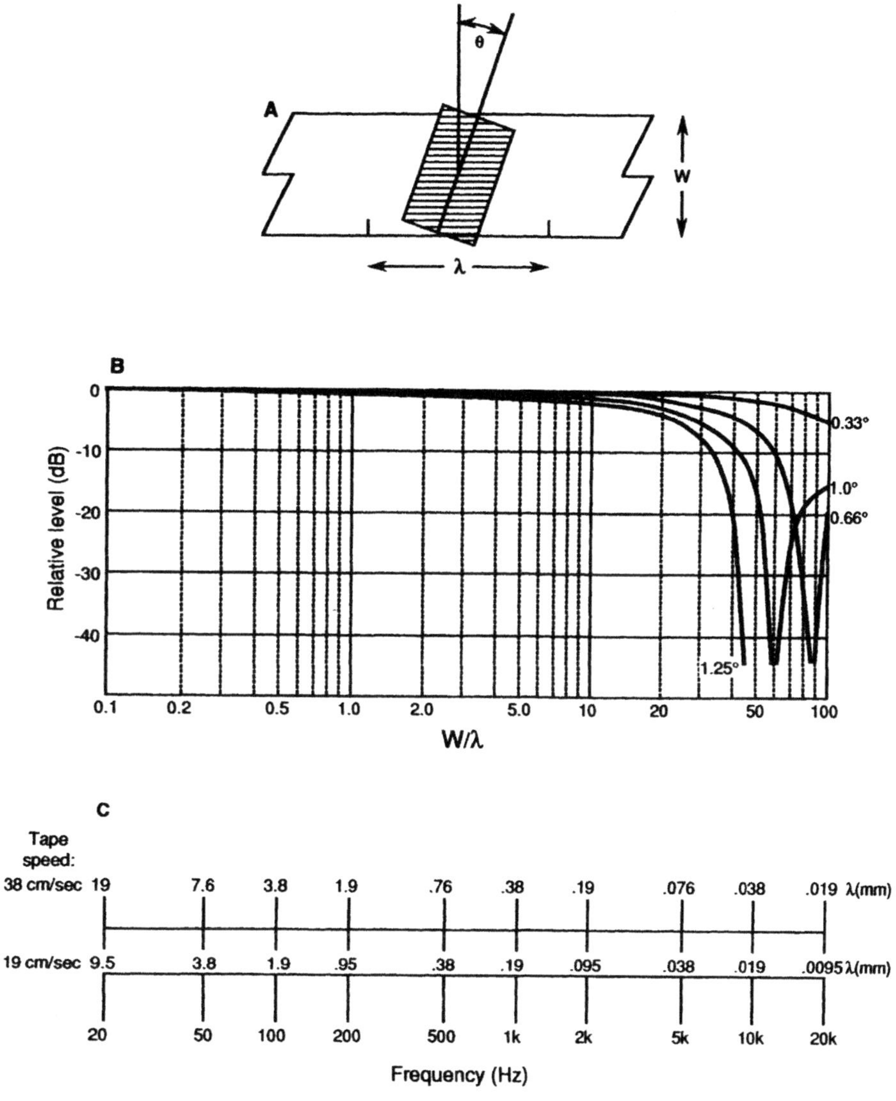

Figure 164. Azimuth Losses in Tape Playback

165. Oxide Thickness Losses in Tape Playback

The thickness of the tape oxide coating layer affects high frequency response and manifests itself as a combination of recording and playback losses. The tradeoff here is that thick oxide coatings promote greater output at long wavelengths at the expense of extended output at shorter wavelengths. The loss is given by the following equation:

$$\text{Response (dB)} = 20 \log \left[\frac{1 - e^{-kt}(1 + kt)}{(kt)^2} \right] \quad \ldots\ldots\ldots\ldots \text{(165.1)}$$

where t is the thickness of the oxide layer and k is the value $2\pi/\lambda$.

EXAMPLE: For an oxide thickness of .006 mm operating at a tape speed of 38 cm/sec and at a frequency of 10 kHz, determine the response loss due to the oxide thickness.

First, determine the value k from the nomograph in Figure 165A. By inspection, this value is 165.

Then, multiply the value k by the oxide thickness .006 mm to obtain the value 0.99. Enter the value of 0.99 along the abscissa; move upward to the bold curve and read the loss of about 12 dB along the ordinate.

References: Bertram (1975)
McKnight (1960)

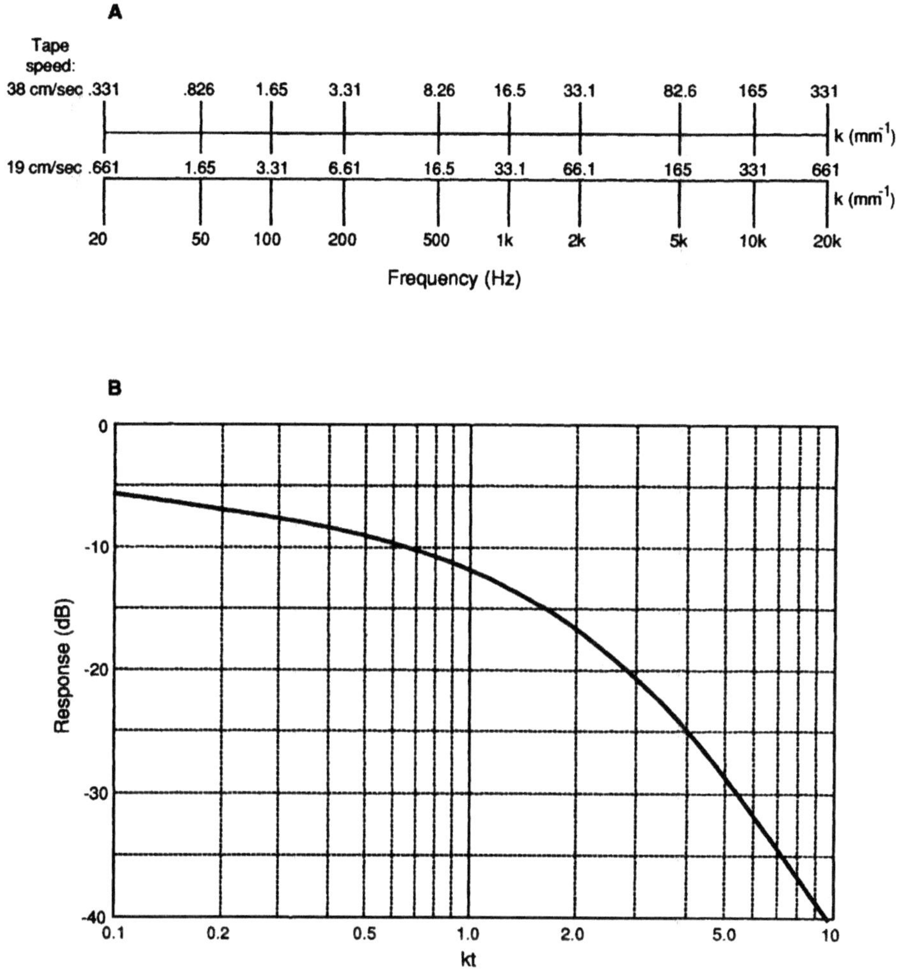

Figure 165. Oxide Thickness Losses in Tape Playback

166. SPACING LOSSES IN TAPE PLAYBACK

Even slight separation of the tape from the head will result in significant losses at short wavelengths. This loss is given by the following equation:

$$\text{Loss (dB)} = 20 \log [\exp (-2\pi d/\lambda)] \quad \text{(166.1)}$$

where d is the average separation of oxide particles from the head surface and λ is the recorded wavelength, both expressed in the same units.

EXAMPLE: For an average oxide-to-head surface spacing of 0.01 mm, a tape speed of 38 cm/sec, and a recorded frequency of 10 kHz, determine the loss due to the spacing.

First, determine the wavelength from the nomograph at Figure 166A. By inspection, this value is .038 mm.

Then, determine $d/\lambda = .01/.038 = 0.26$. On the graph in Figure 166B, locate the value .26 along the abscissa and move upward to the bold curve. The loss is about 14 dB.

References: Busby (1988)
Camras (1988)
Eargle (1992)

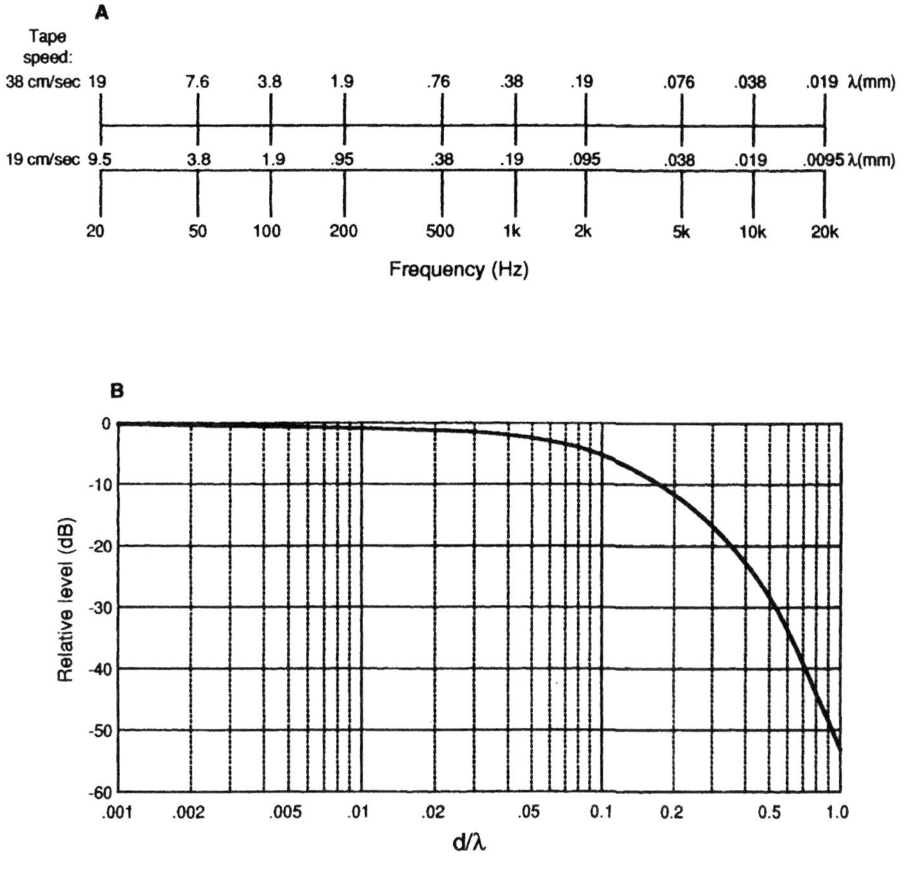

Figure 166. Spacing Losses in Tape Playback

167. Gap Length Losses in Tape Playback

At short wavelengths, the finite size of the playback gap dimension in the direction of tape motion influences the output. This loss is given by the following equation:

$$\text{Response (dB)} = 20 \log \left[\frac{\sin [1.11\pi (g/\lambda)]}{[1.11\pi (g/\lambda)]} \right] \quad \ldots\ldots\ldots\ldots (167.1)$$

where g is the gap length and λ is the recorded wavelength, both expressed in the same units. (The sine term in the numerator must be evaluated in radians.)

EXAMPLE: For a gap length of .006 mm, a tape speed of 19 mm/sec, and a recorded frequency of 20 kHz, determine the gap length loss.

First, determine the wavelength from the nomograph Figure 167A. By inspection, this value is .0095 mm.

Then, determine $g/\lambda = .006/.0095 = .63$. On the graph at Figure 167B, locate the value .63 along the abscissa and move upward to the bold curve. The loss is in the range of 8 dB.

References: Busby (1988)
Camras (1988)
Eargle (1992)

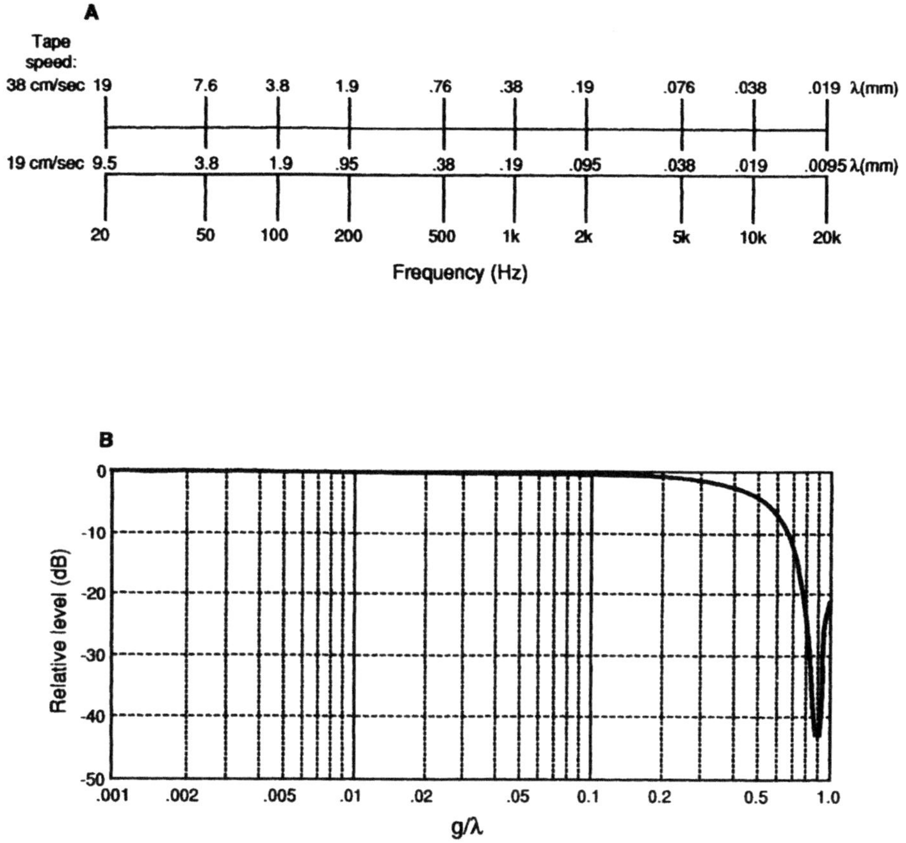

Figure 167. Gap Length Losses in Tape Playback

168. Reference Surface Fluxivity Standards for Tape Recording

The surface fluxivity standards given in Figure 168 are used in the field to establish operating levels for analog magnetic tape recording and playback. For many years, "Ampex reference level" was established by the recorded signal at 700 Hz that produced a given amount of distortion on a given sample of tape stock. In time, absolute measurement of tape surface fluxivity determined the recorded signal to be 185 nanowebers/meters (nW/m).

At the same time, the European recording community had set a reference surface fluxivity of 360 nW/m, which was 5.8 dB higher in level. As discussed in section 100, the near 6-dB difference between the two reference fluxivities was nearly matched by the observed lead factor between the VU meter and peak program meters.

During the 1970s, new magnetic tape was developed that provided higher output capability with no increase in distortion. It became standard practice to elevate the reference level 3 dB so that the added output capability could be conveniently implemented in the studio. As a result, a new generation of elevated test tapes was developed to accommodate the new operating levels. The U.S. standard of 250 nW/m and the European standard of 510 nW/m were adopted for this purpose.

Reference: Eargle (1992)

Surface fluxivity (nWb/m)	Description	Level (dB)	Calibration frequency (Hz)
185	Old Ampex reference level	0	700 Hz
200	Rationalized reference level	+0.7	1 kHz
250	New U. S. elevated level	+2.6	1 kHz
360	Old European (DIN) level	+5.8	1 kHz
510	New European (DIN) level	+8.8	1 kHz

Figure 168. Reference Surface Fluxivity Standards for Tape Recording

169. IEC (International Electrotechnical Commission) Equalization Standards for Professional Tape Playback

Figure 169 shows the response for a flat electrical signal applied to the input of a tape playback preamplifier. For 19 cm/sec playback, the high-frequency boost is greater, corresponding to a time constant of 70 μsec. For 38 cm/sec playback, the high-frequency boost is determined by a 35 μsec boost. The electrical response extends below the range shown here, and the playback curve does not deviate from the 6 dB/octave slope shown. Response is normally continued down to 20 Hz. The high-frequency break frequencies are at 2270 Hz and 4550 Hz.

Because the output of a magnetic playback head is proportional to the rate of change of flux on the tape, a constant flux recording produces flat response with these curves. Rises in high-frequency response commence at the frequencies determined by the time constants.

Because the frequency breakpoints are one octave apart, and since they correspond to two-to-one ratios in tape speed, the IEC curves result in approximately equal amounts of high-frequency boost in recording for either the slower or faster speed.

The same playback curves are used for magnetic film playback; the 35-μsec curve is used for 35-mm and the 70-μsec curve for 16-mm playback.

The equalization used in the record function is very much tape stock– and bias current-dependent. It is customary to equalize the record function to whatever degree is required in order to achieve flat overall response when the curves given here define the playback process.

Reference: Camras (1988)

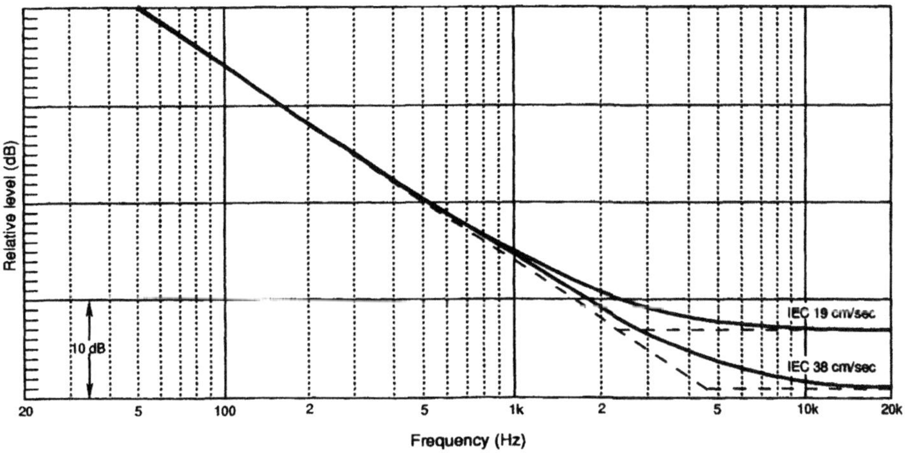

Figure 169. IEC Equalization Standards for Professional Tape Playback

170. NAB (National Association of Broadcasters) Standard for Professional Tape Playback

As magnetic recording gained in popularity just after World War II, the National Association of Broadcasters (NAB) in the United States standardized tape playback equalization for 39 cm/sec. Figure 170 shows the NAB curve. The time constants employed are 50 μsec at high frequencies and 3180 μsec at low frequencies. These numbers correspond to transition frequencies, respectively, of 3180 Hz and 50 Hz.

Later, the same playback curve was adopted for recordings made at 19 cm/sec, requiring that substantially more record equalization be employed at high frequencies compared with 38 cm/sec.

The transition frequency at 50 Hz requires a low-frequency boost in recording in order to produce flat low-frequency response. In principle, this has always disturbed many engineers. There is no good reason for it from the point of view of optimizing the overall record function, and it can result in some low-frequency overload, at least as compared with the IEC standard.

Reference: Camras (1988)

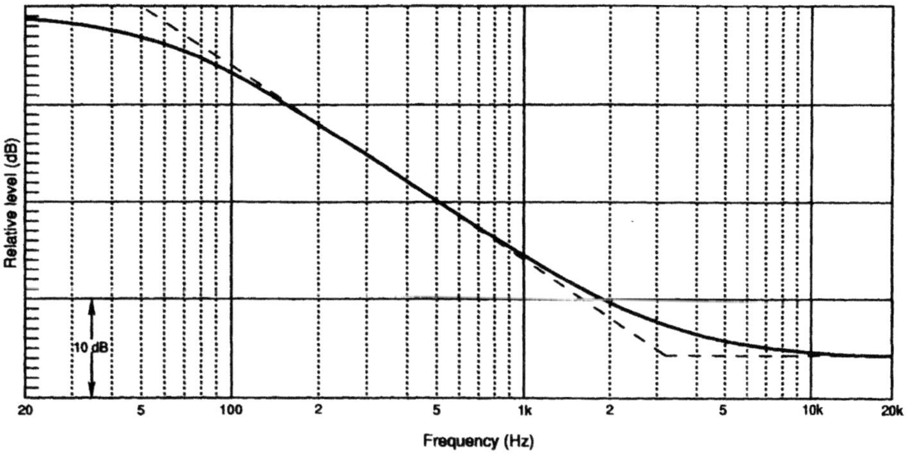

Figure 170. NAB Standard for Professional Playback

171. AES (Audio Engineering Society) Standard for Professional Tape Playback at 76 cm/sec (30 in/sec)

The Audio Engineering Society (AES) curve, shown in Figure 171, was developed during the 1970s. It is virtually an extension of the IEC family of curves. The relevant time constant is 17.5 μsec, corresponding to a high-frequency transition of 9100 Hz.

Its use has normally been in conjunction with new mastering tape formats developed during the 1980s and 1990s to extend the dynamic range of analog tape recording.

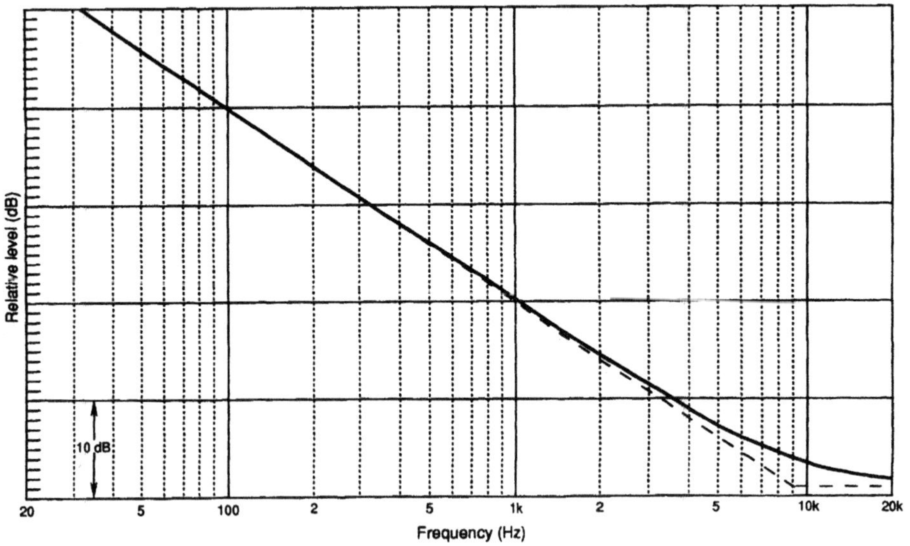

Figure 171. AES Standard for Professional Tape Playback at 76 cm/sec (30 in/sec)

172. STANDARDS FOR PLAYBACK OF CONSUMER TAPE FORMATS

The curves shown in Figure 172 are used in the following consumer formats:

Format	*High-frequency time constant*
9.50 cm/sec (3.750 in/sec)	90 μsec
4.76 cm/sec (1.875 in/sec)	120 μsec
4.76 cm/sec (1.875 in/sec) (high-resolution)	70 μsec

All of these formats have a 3180 μsec time constant at low frequencies.

Reference: Camras (1988)

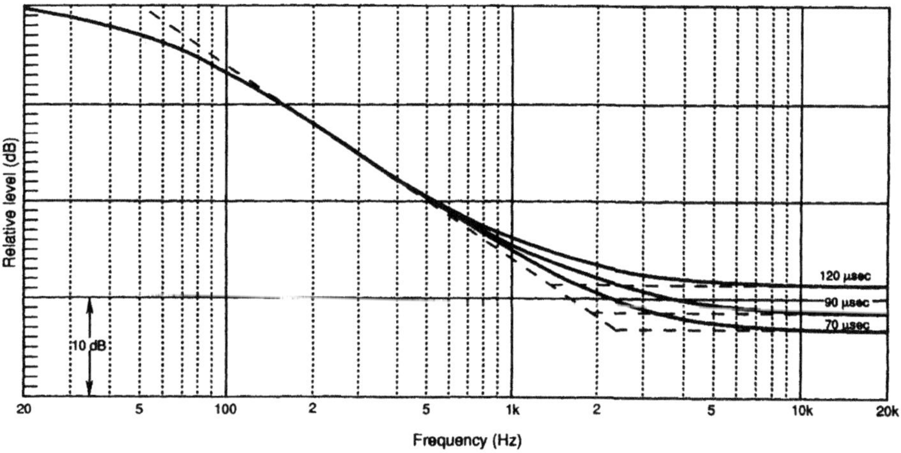

Figure 172. Standards for Playback of Consumer Tape Formats

173. IEC TO NAB CONVERSION AT 38 CM/SEC

On many occasions involving tape exchange between the United States and Europe, it becomes necessary to play an IEC tape on a machine that has only NAB equalization, or vice versa.

The equalization shown at A in Figure 173 is necessary; it must be added external to the tape playback if a 38 cm/sec IEC tape is to be played back on an NAB machine. Note that low frequencies are boosted below 50 Hz and that high frequencies are shelved downward above 2 kHz, as indicated.

If the process is reversed and a 38 cm/sec NAB tape is to be played back on an IEC machine, use curve B.

Reference: Eargle (1992)

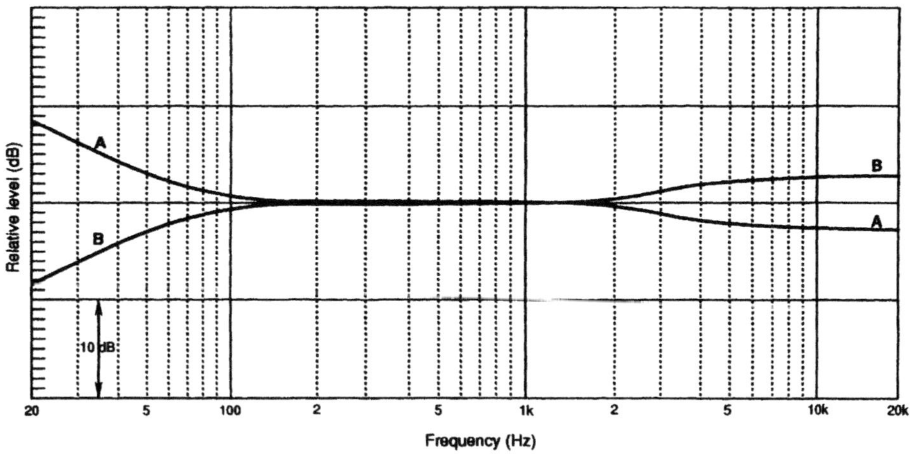

Figure 173. IEC to NAB Conversion at 38 cm/sec

174. IEC TO NAB CONVERSION AT 19 CM/SEC

If you wish to play a 19 cm/sec IEC tape back on a machine equalized for NAB playback, the equalization shown at A in Figure 174 must be added to the signal.

If the process is reversed and a 19 cm/sec NAB tape is to be played back on an IEC machine, use curve B.

Reference: Eargle (1992)

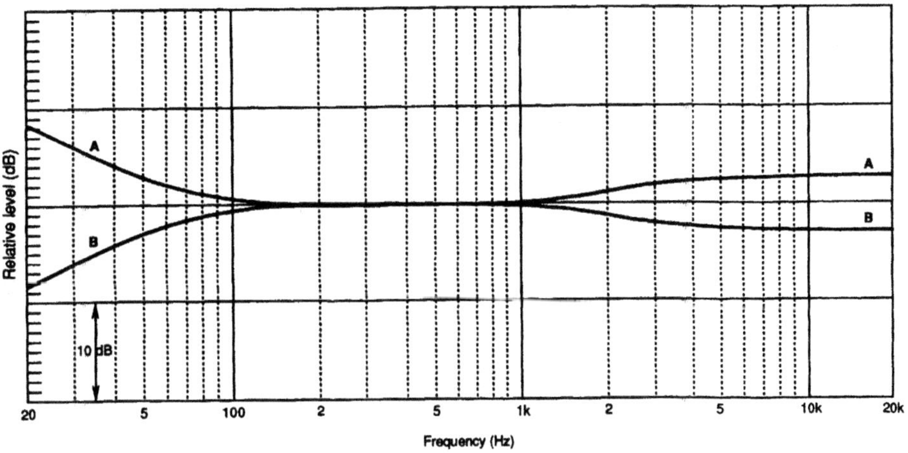

Figure 174. IEC to NAB Conversion at 19 cm/sec

175. Standard Weighting Curve for Tape Flutter Measurements

To make a flutter measurement of a tape recorder, begin by recording a tone of 3 kHz. On playback, the tone is fed to an FM demodulator whose output will be a signal voltage that is proportional to the instantaneous frequency deviation of the 3-kHz tone.

To determine the annoyance effects of flutter, it is customary to use a weighting curve, as shown in Figure 175, ahead of metering the demodulated signal. Because the ear is especially sensitive to flutter components in the 4–5 Hz range, the required filter gives peak output at those frequencies, while rolling off at frequencies above and below.

Reference: Camras (1988)

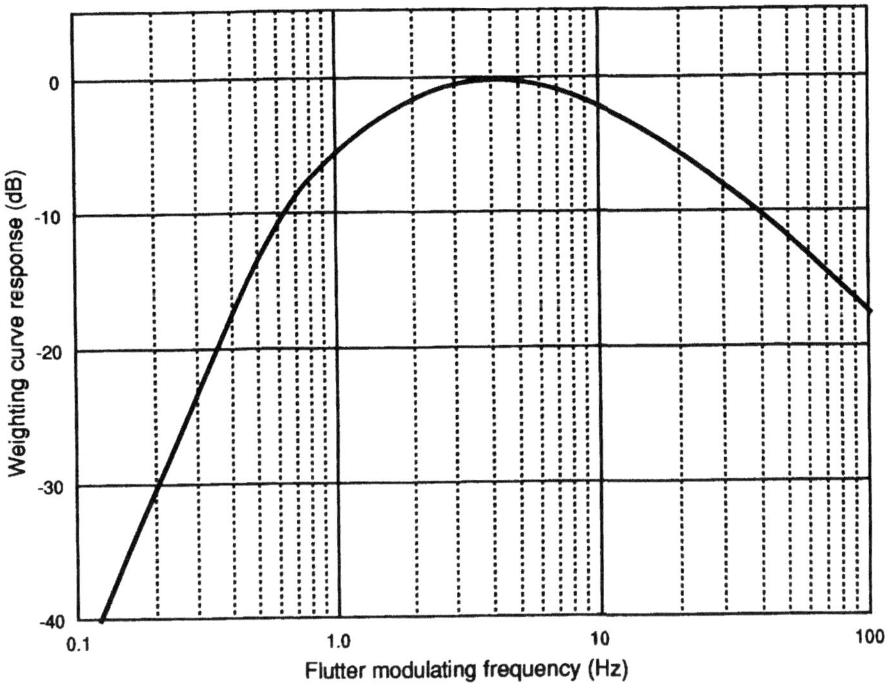

Figure 175. Standard Weighting Curve for Tape Flutter Measurements

UNIT CONVERSION TABLE

	Meter	Kilometer	Inch	Foot	Yard	Mile
Length	1	0.001	39.3701	3.28084	1.09361	0.00062
	1000	1	39370.1	3280.84	1093.61	0.62137
	0.0254	0.00003	1	0.08333	0.02778	0.00002
	0.3048	0.00030	12	1	0.33333	0.00019
	0.9144	0.00091	36	3	1	0.00057
	1609.34	1.60934	63360	5280	1760	1

	Sq Meter	Are	Hectare	Sq Kilometer	Sq Yard	Acre
Area	1	0.01	0.00010	—	1.19599	0.00025
	100	1	0.01	0.00010	119.599	0.02471
	10,000	1000	1	0.01	11960	2.47105
	—	10,000	100	1	—	247.105
	0.83613	0.00836	0.00004	—	1	0.00021
	4046.86	40.4686	0.40469	0.00405	4840	1

	Cu Meter	Liter	U.S. Gallon	Cu Inch	Cu Foot	Imperial Gallon
Volume	1	1000	264.171	61023.7	35.3147	219.978
	0.001	1	0.26418	61.0237	0.03532	0.21998
	0.00379	3.78532	1	231.001	0.13368	0.83270
	0.00002	0.01639	0.00433	1	0.00058	0.00360
	0.02832	28.316	7.48048	1728	1	6.22883
	0.00455	4.54596	1.20095	277.42	0.16054	1

	Gram	Kilogram	Metric Ton	Carat	Ounce	Pound
Weight	1	0.00100	—	5.00000	0.03527	0.00220
	1000.00	1	0.001	5000.00	35.2740	2.20462
	—	1000.00	1	—	35274.0	2204.62
	0.2	0.0002	—	1	0.00705	0.00044
	28.3495	0.02835	0.00003	141.748	1	0.06250
	453.592	0.45359	0.00045	2267.95	16	1

REFERENCES

Augspurger, G. 1971. Elertrical versus acoustical parameters in the design of loudspeaker crossover networks. *J. Audio Engineering Soc.* 19(b).

Backus J. 1969. *The Acoustical Foundations of Music.* New York: W.W. Norton.

Barron, M. 1971. The subjective effects of first reflections in concert halls. The need for lateral reflections. *J. Sound Vibration* 15:475–94.

Bauer, B. 1940. Graphical determination of the random efficiency of microphones. *J. Acoust. Soc. America* 11:477–9.

———. 1956. Phasor analysis of some stereophonic phenomena. *J. Acoust. Soc. America* 33:(11).

Beavers, B., and R. Brown. 1970. Third-order gradient microphone for speech reception. *J. Audio Engineering Soc.* 18(6).

Beers, G., and H. Belar. 1943. Frequency modulation distortion in loudspeakers. *J. Audio Engineering Soc.* 29(5).

Benade, A. 1972. *Fundamentals of Musical Acoustics.* New York Oxford Univ. Press (reprinted New York: Dover, 1990).

Beranek, L. 1954. *Acoustics.* New York: McGraw Hill.

———. 1971. *Noise and Vibration Control.* New York: McGraw Hill.

Bertram, H. 1975. Wavelength response in ac biased recording. *IEEE Transactions on Magnetics.* 11(5).

Blauert, J. 1983. *Spatial Hearing.* Cambridge, MA: MIT Press.

Blauert, J., and P. Laws. 1978. Group delay distortion in electroacoustical systems. *J. Acoust. Soc. America.* 63(5).

Bolt and Doak. (1950). A tentative criterion for the short term transient response of auditoriums. *J. Acoust. Soc. America* 22:507–9.

Bore, G. 1989. *Microphones,* G. Neumann, GMBH, Berlin.

Brüel and Kjær. 1971. *Acoustic noise measurements.* Denmark: B & K.

Busby, E. 1988. Analog magnetic tape recording and reproduction. In: Benson, *Audio Engineering Handbook.* New York: McGraw Hill.

Camras, M. 1988. *Magnetic Recording Handbook.* New York: Van Nostrand Reinhold.

Ceoen, C. 1972. Comparative stereophonic listening tests. *J. Audio Engineering Soc.* 20:19–27.

Churcher and King 1937. *J. Inst. Electr. Engineers* 81:57–90.

Clark, H., et al. 1958. The "stereosonic" recording and reproducing system. *J. Audio Engineering Soc.* 6(2).

Cohen, E., and L. Fielder. 1992. Determining noise criteria for recording environments. *J. Audio Engineering Soc.* 40(5).

Cremer, L., and H. Müller. 1978. *Principles and Applications of Room Acoustics, I.* New York: Applied Science Publishers.

Damaske, P. 1967. Subjective investigation of sound fields. *Acustica* 22(3).

Davis, D., and C. Davis. 1987. *Sound System Engineering.* Indianapolis, IN: Howard W. Sams.

Dolby Laboratories. 1993. *Technical Guidelines for Dolby Stereo Theaters.* San Francisco, CA: Dolby Laboratories Incorporated.

Dooley, W., and R. Streicher. 1985. MS stereo: A powerful technique for working in stereo. *J. Audio Engineering Soc.* 33(7/8).

Eargle, J. 1981. *The Microphone Handbook.* Commack, NY: ELAR Publishing.

———. 1989. *Handbook of Sound System Design.* Commack, NY: ELAR Publishing.

———. 1990. *Music, Sound, and Technology.* New York: Van Nostrand Reinhold.

———. 1992. *Handbook of Recording Engineering.* Van Nostrand Reinhold, New York.

Eargle, J., J. Bonner, and D. Ross. 1985. The academy's new state of the art loudspeaker system. *J. Soc. Motion Picture Television Engineers* 94(6).

Engebretson, M. 1984. Low-frequency sound reproduction. *J. Audio Engineering Soc.* 32:340–6.

Faulkner, T. 1981. A phased array. *Hi-Fi News and Record Review* (July 1981).

Fletcher, H., and W. Munson. 1933. Loudness, its definition, measurement, and calculation. *J. Acoust. Soc. America* 5:82–108.

Fletcher, N., and T. Rossing. 1991. *The Physics of Musical Instruments.* New York: Springer-Verlag.

Foreman, C. 1987. Sound system design. In: Ballou, G. *Handbook for Sound Engineers.* Indianapolis, IN: Howard W. Sams.

Franssen, N. V. 1963. *Stereophony.* Eindhoven, Netherlands: Philips Technical Bibliography.

Frayne, J., and H. Wolfe. 1949. *Elements of Sound Recording.* New York: Wiley.

French, N., and J. Steinberg. 1947. Factors governing the intelligibility of speech sounds. *J. Acoust. Soc. America.* 19:90–119.

Gander, M. 1981. Moving-coil loudspeaker topology as an indicator of linear excursion capability. *J. Audio Engineering Soc.* 29:10–26.

Gander, M., and J. Eargle. 1990. Measurement and estimation of large loudspeaker array performance. *J. Audio Engineering Soc.* 38:204–20.

Glover, R. 1940. A review of cardioid type unidirectional microphones. *J. Acoust. Soc. America.* 11:296–302.

Greiner, R., and J. Eggers. 1989. The spectral amplitude distribution of selected compact discs. *J. Audio Engineering Soc.* 37(4).

Haas, H. 1972. The influence of a single echo on the audibility of speech. *J. Audio Engineering Soc.* 20(2).

Harris, C. 1979. *Handbook of Noise Control.* New York: McGraw-Hill.

Hibbing, M. 1989. XY and MS microphone techniques in comparison. *Sennheiser News* (Wedemark, F.R.G.: Sennheiser Electric Corporation).

Holman, T. 1988. Postproduction systems and editing. In: Benson, *Audio Engineering Handbook.* New York: McGraw-Hill.

JBL. 1982. *Sound System Design Reference Manual.* Northridge, CA: JBL Incorporated.

———. 1984. JBL *Technical note,* 1(7). Northridge, CA: JBL Incorporated.

Keele, D. 1977. Low frequency horn design using Thiele-Small driver parameters. Presented at the Audio Engineering Society Convention, Los Angeles, CA. May 1977; preprint number 1250.

Klepper, D., and D. Steele. 1963. Constant directional characteristics from a line source array. *J. Audio Engineering Soc.* 11(3).

Kryter, K. 1962. Methods for the calculation and use of the articulation index. *J. Acoust. Soc. America* 34:1689.

Kuttruff, H. 1973. *Room Acoustics.* New York: Applied Science Publishers.

———. 1979. *Room Acoustics.* London: Applied Science Publishers.

Locanthi, B. 1952. Application of electric circuit analogies to loudspeaker design problems. *IRE Transactions on Audio, vol.* PGA-1.

Lubman, D., and E. Wetherill. 1985. *Acoustics of Worship Spaces.* New York: Acoustical Society of America.

Massa, F. 1942. *Acoustic Design Charts.* Philadelphia: Blakiston.

McKnight, J., 1960. The Frequency Response of Magnetic Recorders for Audio. *J. Audio Engineering Society,* 8(3).

Metzler, B. 1993. *Audio Measurement Handbook.* Beaverton, OR: Audio Precision Incorporated.

Meyer, J. 1978. *Acoustics and the Performance of Music.* Frankfurt/Main: Verlag das Musikinstrument.

Molloy, C. 1948. Calculation of the directivity index for various types of radiators. *J. Acoust. Soc. America* 20:387–405.

Muller, X., X. Dunn, and Black. 1938. *J. Acoust. Soc. America* 10:6.

Newitt, J. 1953. *High Fidelity Techniques.* New York: Rinehart.

Olson, H. 1957. *Acoustical Engineering.* New York: D. Van Nostrand.

Olson, L. 1979. The stereo-180 microphone system. *J. Audio Engineering Soc.* 27:158–63.

Peutz, V. 1971. Articulation loss of consonants as a criterion for speech transmission in a room. *J. Audio Engineering Soc.* 19(11).

———. 1973. Quasi-steadystate and decaying sound fields. *Ingenieursblad* 42(8) (in Dutch).

Pollack, X. 1952. *J. Acoust. Soc. America* 24:533–8.

Rathe, E. 1969. Notes on two common problems of sound propagation. *J. Sound Vibration* 10:472–9.

Robinson, D., and R. Dadson. 1956. A re-determination of the equal loudness relations for pure tones. *Br. J. Appl. Physics* 7:166–81.

Roederer, J. 1973. *Introduction to the Physics and Psychophysics of Music.* New York: Springer-Verlag.

Rossing, T. 1990. *The Science of Sound.* Reading, MA: Addison-Wesley.

Sank, J. 1985. Microphones. *J. Audio Engineering Soc.* 33(7/8).

Schultz, T. 1984. Paper given at Acoustical Society of America meeting.

Sepmeyer, L. 1988. Measurement, direct field bias, and critical frequency in the measurement of reverberant fields. *J. Acous. Soc. America.* 83(4).

Small, R. 1973. Vented box loudspeaker systems, part III. *J. Audio Engineering Soc.* 21(7).

Smith, H. 1981. Acoustic design considerations for speech intelligibility. *J. Audio Engineering Soc.* 29(6).

Spoor, A. 1967. *Int. Audiology* 6:48.

Staffeldt, H. 1984. Measurement and prediction of the timbre of sound reproduction. *J. Audio Engineering Soc.* 32(6).

Stevens, S. 1935. The relation of pitch to intensity. *J. Acoust. Soc. America* 6:150–4.

Stevens, S., and J. Volkmann. 1940. The relation of pitch to frequency: A revised scale. *Am. J. Psychol.* 53:329–53.

Taleske, R., E. Wetherill, and W. Cavanaugh. 1982. *Halls for Music Performance: Two Decades of Experience: 1962–1982.* New York: Acoustical Society of America.

Terhardt, E. 1979. Calculating Virtual Pitch. *Hearing Res.* 1:155.

Thiele, N. 1985. Three-level tone signal for setting audio levels. *J. Audio Engineering Soc.* 33(12).

Watson, C., and R. Gengel. 1969. Signal duration and signal frequency in relation to auditory sensitivity. *J. Acoust. Soc. America* 46:989–97.

Western Electro-Acoustic Laboratory. (1959). Study and investigation of specialized transducers for voice communication. U.S. Air Force Contract AF33(616)-3710, Task no. 43060. Santa Monica, CA: Western Electro-Acoustic Laboratory.

Wolff, I., and L. Malter. 1929. Sound radiation from a system of circular diaphragms. *Phys. Rev.* 13:1061.

Woram, J. 1989. *Sound Recording Handbook.* Indianapolis, IN: H. W. Sams.

INDEX

GPSR Compliance

The European Union's (EU) General Product Safety Regulation (GPSR) is a set of rules that requires consumer products to be safe and our obligations to ensure this.

If you have any concerns about our products, you can contact us on ProductSafety@springernature.com

In case Publisher is established outside the EU, the EU authorized representative is:

Springer Nature Customer Service Center GmbH
Europaplatz 3
69115 Heidelberg, Germany

Batch number: 10755284

Printed by Printforce, the Netherlands